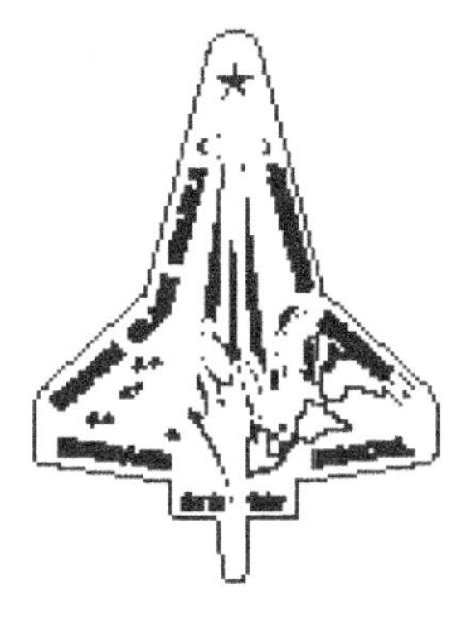

As we celebrate the "Centennial of Flight," let us remember the contributions of the Columbia STS-107 crew members.

Spinoff salutes the STS-107 crew who "dedicated their lives to pushing scientific challenges for all of us here on Earth. They dedicated themselves to that objective and did it with a happy heart, willingly and with great enthusiasm..."

—NASA Administrator Sean O'Keefe, February 1, 2003

On the Cover:

From the first wind tunnel to the latest aircraft and spacecraft designs, the montage displays several of the many contributions made by NASA and its predecessor, NACA, during the "100 Years of Powered Flight."

Spinoff 2003

National Aeronautics and Space Administration
Office of Aerospace Technology
Commercial Technology Division

Developed by
Publications and Graphics Department
NASA Center for AeroSpace Information (CASI)

National Aeronautics and
Space Administration

NASA
www.nasa.gov
The NASA Vision
To improve life here,
To extend life to there,
To find life beyond.
The NASA Mission
To understand and protect our home planet,
To explore the universe and search for life,
To inspire the next generation of explorers
... as only NASA can.

Foreword

In this "Centennial of Flight" year, it is worth noting for all the amazing progress the age of aviation and space flight has made possible, our gains have not come easily. Every step of the way, the technological breakthroughs that have enabled people to fly around the world and brave explorers to extend our horizons heavenward, were the result of hard work, perseverance, and a willingness to overcome major setbacks.

On February 1, 2003, a terrible tragedy occurred when the NASA family, our Nation, and the world lost seven remarkable individuals, the heroic crew of the Space Shuttle Columbia.

NASA is now working hard to return to space flight operations that are as safe as humanly possible. We are also continuing to pursue our mission goals of understanding and protecting the home planet, exploring the universe and searching for life, and inspiring the next generation of explorers. We hope our unceasing efforts to pioneer the future will provide a fitting tribute to the Columbia seven.

This year, which marks NASA's 45th year of conducting aeronautics and space research and exploration missions on behalf of the American public, is also noteworthy for some important advances:

- Our Expedition crews onboard the International Space Station continued to perform experiments on the orbiting facility spanning several scientific disciplines. From these experiments, scientists are: learning better methods of drug testing; developing models that predict or explain the progress of disease; investigating how to use microbes to make antibiotics; determining how to improve manufacturing processes; and studying changes in Earth climate, vegetation, and crops.
- We successfully launched our twin Mars Exploration Rovers, Spirit and Opportunity, which are now en route for their January 2004 exploration of sites on the Red Planet where water may have once flowed freely.
- We celebrated the Nobel Prize in Physics awarded in December 2002 to astronomer Riccardo Giacconi for his groundbreaking NASA-sponsored research in X-ray astronomy.
- We initiated the NASA Explorer School Program to provide fifth through eighth grade-level educators, administrators, students, and their families the opportunity to engage in sustained involvement with NASA's research, discoveries, and missions. NASA also began a program to recruit the first class of Educator Astronauts, who in addition to performing regular flight duties on multiple missions, will take their classrooms into space to directly engage millions of school children in lessons about the wonders of science.
- We launched several satellites and instruments that are helping scientists better understand the dynamics of Earth's climatic system and the possible causes and consequences of global climate change.

Fittingly for an Agency that holds dear its aviation roots, we also continued to make significant investments to improve the efficiency, safety, and security of our Nation's air transportation system. *Spinoff 2003* recognizes a number of exciting NASA aeronautics research efforts that may well help revolutionize the way we travel in the future.

As always, this publication highlights NASA's extensive efforts to promote the transfer of aerospace technology to the private sector. Every day, in an astounding variety of ways, American lives are affected positively by our Nation's investment in NASA. Such fields as agriculture, communications, computer technology, environment and resources management, health and medicine, manufacturing, transportation, and climate modeling have benefited greatly from NASA-derived technologies.

As the second century of flight gets underway, we will strive mightily to continue providing tangible and significant benefits to the American public. For at NASA, we believe the sky and the heavens beyond are not limits, but rather vital venues for exploration and technological progress.

Sean O'Keefe
Administrator
National Aeronautics and Space Administration

Introduction

NASA's enduring contributions in aerospace research and development trace their origin to the Wright Brothers' historic flight in December 1903. This year, we celebrate the 100th anniversary of that flight and almost 10 decades of the National Advisory Committee for Aeronautics' (NACA) and NASA's spectacular accomplishments in space and here on Earth. In 1915, Congress established NACA "to supervise and direct the scientific study of the problems of flight, with a view to their practical solution." Congress could not anticipate at that time the future impact this legislation would have for every American and the global community.

Today, NASA continues to reach milestones in space exploration with the Hubble Telescope, Earth-observing systems, the Space Shuttle, the Stardust spacecraft, the Chandra X-Ray Observatory, the International Space Station, the Mars rovers, and experimental research aircraft—these are only a few of the many initiatives that have grown out of NASA engineering know-how to drive the Agency's missions. The technical expertise gained from these programs has transferred into partnerships with academia, industry, and other Federal agencies, ensuring America stays capable and competitive.

With *Spinoff 2003*, we once again highlight the many partnerships with U.S. companies that are fulfilling the 1958 Space Act stipulation that NASA's vast body of scientific and technical knowledge also "benefit mankind." This year's issue showcases innovations such as the cochlear implant in health and medicine, a cockpit weather system in transportation, and a smoke mask benefiting public safety; many other products are featured in these disciplines, as well as in the additional fields of consumer/home/recreation, environment and resources management, computer technology, and industrial productivity/manufactacturing technology.

Also in this issue, we devote an entire section to NASA's history in the field of flight and showcase NASA's newest enterprise dedicated to education. The Education Enterprise will provide unique teaching and learning experiences for students and teachers at all levels in science, technology, engineering, and mathematics. The Agency also is committed, as never before, to engaging parents and families through NASA's educational resources, content, and opportunities. NASA's catalyst to intensify its focus on teaching and learning springs from our mission statement: "to inspire the next generation of explorers ... as only NASA can."

NASA has proven in the past that it is up to the task and that it is ready for the future.

Dr. Robert L. Norwood
Director, Commercial Technology Division
National Aeronautics and Space Administration

Dr. Adena Williams Loston
Associate Administrator for Education
National Aeronautics and Space Administration

Table of Contents

Co
Nb
Ta
Cancel
20.0
10.0

Partnership Benefits

Each year, NASA makes breakthroughs in science and technology that expand our knowledge of Earth and the universe. Through these advances, NASA extends partnership opportunities for private industry to develop innovative products and services for the American consumer. The following partnership benefits, which serve as NASA's portal to the public, demonstrate the ways space research and development strengthen our economy and improve life here on Earth.

The Right Track for Vision Correction

More and more people are putting away their eyeglasses and contact lenses as a result of laser vision correction surgery. LASIK, the most widely performed version of this surgical procedure, improves vision by reshaping the cornea, the clear front surface of the eye, using an excimer laser. One excimer laser system, Alcon's LADARVision® 4000, utilizes a laser radar (LADAR) eye tracking device that gives it unmatched precision.

During LASIK surgery, laser pulses must be accurately placed to reshape the cornea. A challenge to this procedure is the patient's constant eye movement. A person's eyes make small, involuntary movements known as saccadic movements about 100 times per second. Since the saccadic movements will not stop during LASIK surgery, most excimer laser systems use an eye tracking device that measures the movements and guides the placement of the laser beam.

The eye tracking device must be able to sample the eye's position at a rate of at least 1,000 times per second to keep up with the saccadic movements. Eye tracking devices vary greatly in speed depending upon their type. The most commonly used video tracking systems follow eye movements between 60 to 250 times per second, meaning that they cannot keep up with saccadic movements. Therefore, when the eye moves too far from the limits set by the eye surgeon, the surgeon must shut down the laser beam and restart the surgery once the laser is properly centered.

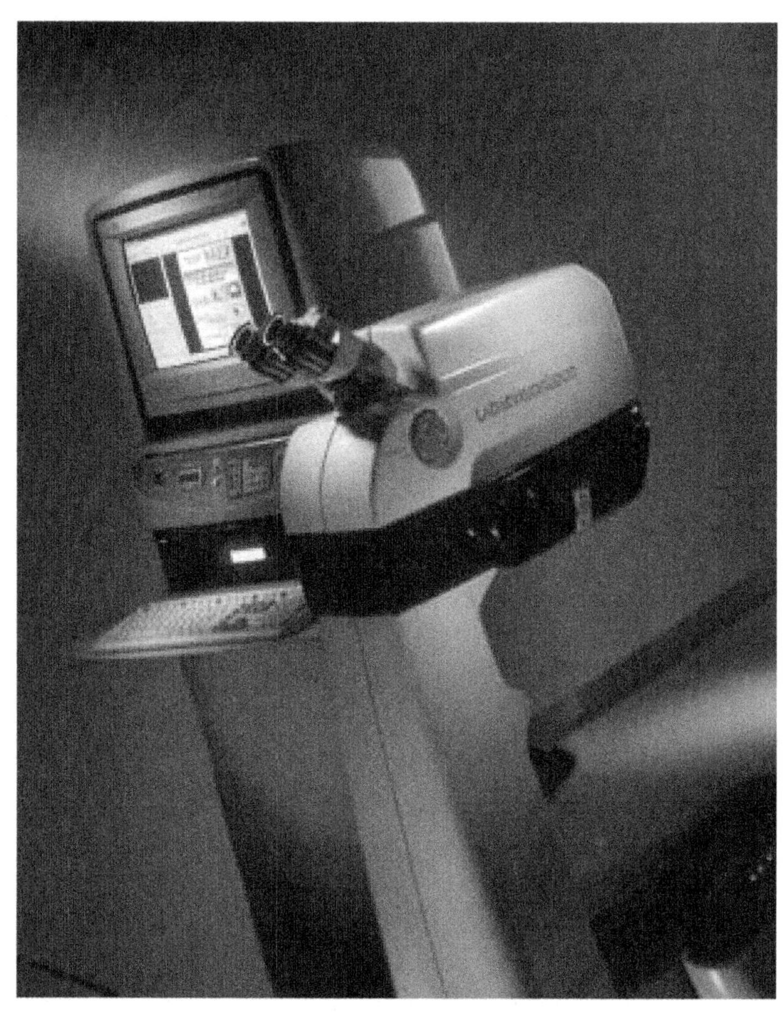

LADARVision® is approved by the U.S. Food and Drug Administration for the correction of nearsightedness, farsightedness, and astigmatism through LASIK surgery.

Sufficient speed is not a problem for Alcon's patented LADARTracker,™ a LADAR eye tracking device that measures eye movements at a rate of 4,000 times per second, 4 times the perceived safety margin. The LADARTracker also employs a closed-loop system, which keeps the device locked on the eye at all times. Eye movement information is continuously relayed to the system, allowing the system to compensate for the movements. Video tracking systems, on the other hand, are open-loop systems that try to follow eye movements rather than compensate for them. LADARVision is currently the only excimer laser device that continually monitors and tracks eye movement through the closed-loop system.

LADARVision's eye tracking device stems from the LADAR technology originally developed through several **Small Business Innovation Research (SBIR)** contracts with NASA's Johnson Space Center and the U.S. Department of Defense's Ballistic Missile Defense Office (BMDO). In the 1980s, Johnson awarded Autonomous Technologies Corporation a Phase I SBIR contract to develop technology for autonomous rendezvous and docking of space vehicles to service satellites. During Phase II of the Johnson SBIR contract, Autonomous Technologies developed a prototype range and velocity imaging LADAR to demonstrate technology that could be used for this purpose. LADAR was also used in military and NASA-sponsored research for applications in strategic target tracking and weapons firing control.

Autonomous Technologies' work for NASA in the area of pointing and scanning laser beams aided the development of the eye tracking device and the LADARVision system. With its advances in LADAR, Autonomous Technologies decided to enter the excimer laser business to develop a system for refractive surgery. In 1998, the U.S. Food and Drug Administration (FDA) granted Autonomous Technologies approval to market its LADARVision system for the correction of nearsightedness, farsightedness, and astigmatism. Shortly afterwards, a subsidiary of Summit Technology, Inc., an industry leader in ophthalmic excimer laser technology, merged with Autonomous Technologies, forming Summit Autonomous.

Alcon acquired Summit Autonomous and the LADARVision technology in May 2000, placing the Fort Worth, Texas-based company at the forefront of refractive surgical technology. The company's LADARVision 4000 made another breakthrough by combining the benefits of the LADAR tracking device with a flying, small-spot laser beam. This small, narrow laser beam is 0.8 millimeters wide, permitting a very precise, gradual corneal shaping. The eye surgeon can closely calibrate the beam to remove the proper amount of corneal tissue for correction of the refractive errors that cause vision problems. The system has the only FDA-approved claim of improved accuracy in corneal shaping.

Eye surgeons across the country are utilizing the LADARVision 4000 for LASIK surgery. In October 2002, Alcon's LADARVision system, consisting of the LADARVision 4000 and the LADARWave wavefront measurement device, became the first to gain FDA approval for wavefront-guided LASIK. The resulting procedure, called CustomCornea, allows surgeons to measure and address visual distortions that previously went undetected. The precision of the tracking device and the small spot beam make the LADARVision system the premier equipment to deliver these precise treatments.

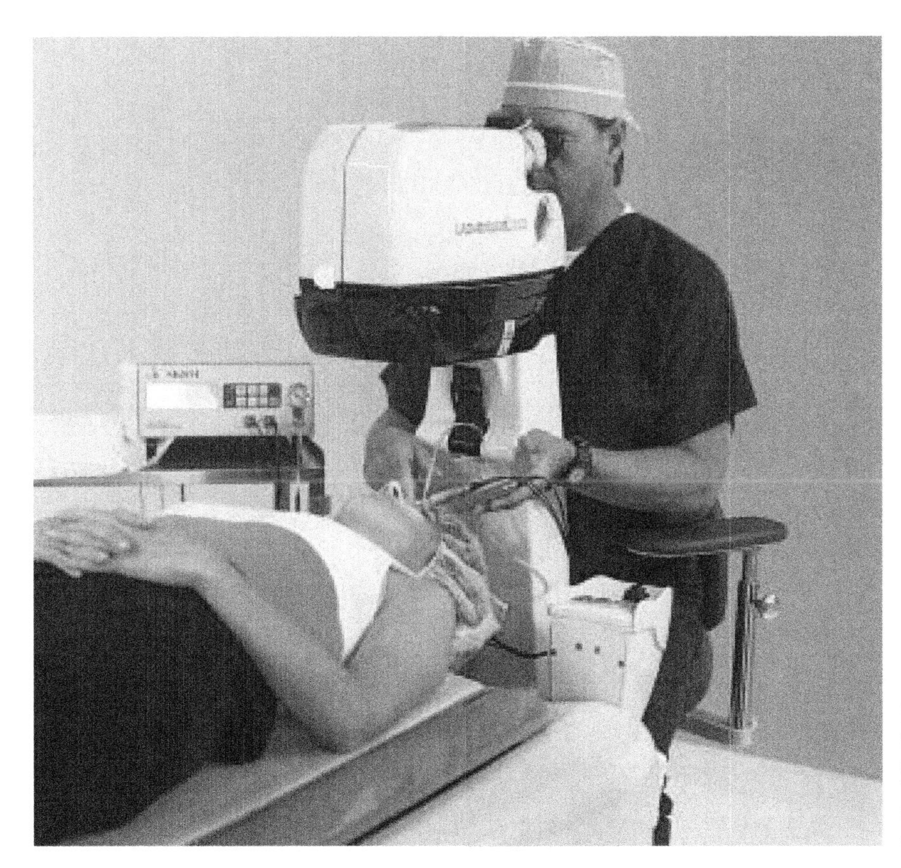

During LASIK surgery, Alcon's LADARTracker™ measures eye movements at a rate of 4,000 times per second. This is 4 times the perceived safety margin for an eye tracking device.

A Real Attention-Getter

While most parents would agree that playing videos games is the antithesis of time well spent for their children, recent advances involving NASA biofeedback technology are proving otherwise.

The same techniques used to measure brain activity in NASA pilots during flight simulation exercises are now a part of a revolutionary video game system that is helping to improve overall mental awareness for Americans of all ages, including those who suffer from Attention Deficit Hyperactivity Disorder (ADHD). For years, scientists from NASA's Langley Research Center have researched and developed various physiological methods for assessing sustained attention, engagement, awareness, and pilot stress in laboratory flight simulators. Such tests are crucial to maintaining the focus of pilots, taking into consideration that the task of flying a plane can sometimes be monotonous.

One of the most progressive physiological methods to spawn from Langley biofeedback research is known as Extended Attention Span Training (EAST). As a modification of biocybernetic technology used to increase the mental engagement of pilots, EAST transcends conventional neurofeedback systems by taking the form of a video game that responds to brain electrical activity and joystick input.

Langley awarded CyberLearning Technology, LLC, of Plymouth Meeting, Pennsylvania, an exclusive license to transform the EAST technology into a fun and exciting video game platform that could safely improve brain functioning for individuals with attention disorders, as well as those who endure high stress and anxiety. In 2003, CyberLearning Technology released the S.M.A.R.T. (Self Mastery and Regulation Training) BrainGames system, an interactive, at-home training tool that is completely compatible with off-the-shelf Sony PlayStation® video games, including such popular titles as Gran Turismo,® Tony Hawk's Pro Skater,™ and Spyro the Dragon.™ The S.M.A.R.T. BrainGames product uses electroencephalogram (EEG) neurofeedback to make a video game respond to the activity of the player's body and brain. Signals from sensors attached to the player's head and body are fed through a signal-processing unit, and then to a video game controller. As the player's brainwaves come closer to an optimal state of attention, the video game's controller becomes easier to control. On the contrary, if a player becomes bored or distracted, the brainwaves stray from the desired stress-free pattern, and controlling the game becomes more difficult. This encourages the player to continue producing optimal patterns or signals to succeed at the game.

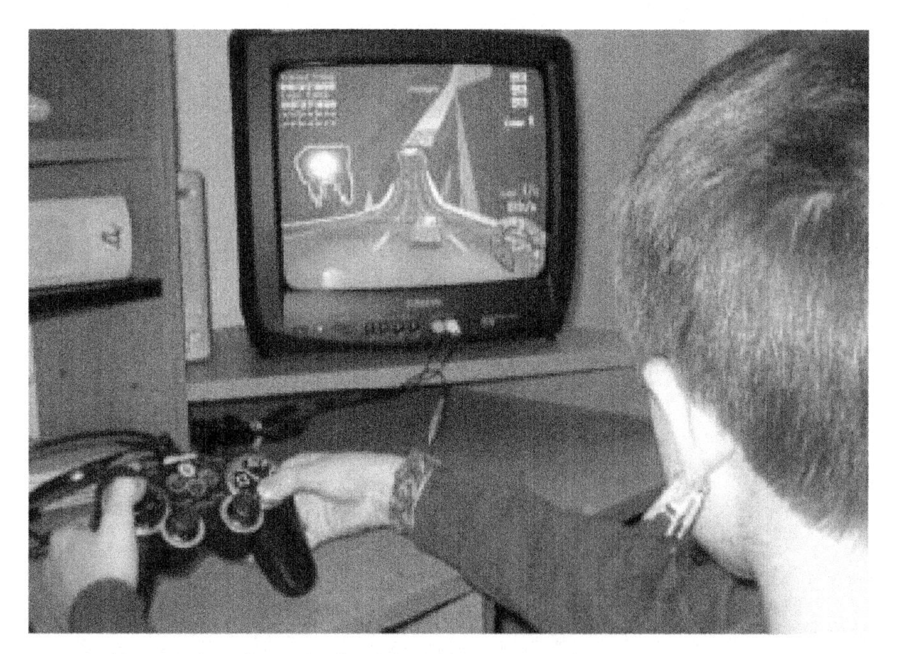

From flight simulation to brain stimulation: The S.M.A.R.T. BrainGames system uses electroencephalogram neurofeedback to make a video game respond to the activity of the player's body and brain.

For example, if an individual is engaging in a race car game in which the goal is to post a fast time to qualify for the next race, it is important for him or her to maintain both speed and control. As the user improves focus, the S.M.A.R.T. BrainGames system will allow for faster speed and easier steering; if the user's focus wanders, the race car will lose ground and not qualify. In essence, the brain acts as the "accelerator," and the calmness acts as the "steering," notes CyberLearning Technology. In the case of ADHD, where relentless distractions and/or impulsive behavior take over, the biofeedback technology behind S.M.A.R.T. BrainGames has displayed great results in helping those with the disorder to concentrate and self-regulate.

Compatible with off-the-shelf Sony PlayStation® video games, the interactive, at-home video training tool fully preserves high-tech entertainment value, unlike previous biofeedback methods that had a propensity to be too repetitive and simplistic.

Researchers from Langley and the Eastern Virginia Medical School in Norfolk conducted a study on the effectiveness of the video game biofeedback compared with traditional biofeedback treatments on 22 boys and girls between the ages of 9 and 14. In the test, six PlayStation games were used; one-half of the children received traditional biofeedback training, and the other half played the modified video games.

After forty 1-hour sessions, both groups showed significant improvements in everyday brain-wave patterns, as well as in tests measuring attention span, impulsiveness, and hyperactivity. The key difference in the outcome, however, was motivation. According to Alan Pope, Ph.D., a psychologist from Langley's Crew/Vehicle Integration Branch and co-inventor of EAST, the video game group experienced fewer no-shows for testing and no drop-outs. Additionally, Pope adds that the parents were more satisfied with the results of the video game training, and the kids seemed to have more fun. He is also quick to note that violent video games are not recommended, but rather that car racing, skateboarding, and other skills-type games are best suited for the interactive technology. By adapting to today's most popular video games, S.M.A.R.T. BrainGames fully preserves high-tech entertainment value, unlike previous biofeedback methods that had a propensity to be too repetitive and simplistic. These training methods typically employed "go-no-go" games, in which animation or computer graphics would move along a predetermined path, lacking interactivity or user control over the game itself.

S.M.A.R.T. BrainGames' motivating and mind-expanding capabilities are also helping to deflect parental criticism regarding the negative influences of video games, such as their ability to keep children away from their homework, or from outdoor playtime activities that are valuable to their social development (CyberLearning Technology's clever answer to such concern is that it is now okay to tell kids to "go play your video games before your homework"). More so, the game system is a viable alternative to frontline medicine for ADHD patients, such as the stimulant Ritalin. Although Ritalin treatment has had great success in controlling the symptoms of ADHD, physicians generally agree that the drug is over-prescribed. CyberLearning Technology stresses that the S.M.A.R.T. BrainGames product should be viewed as an adjunct treatment to medicine, not as a competitor.

Pope notes that this spinoff could have "spin-back" applications for NASA. The Agency has future plans to use the video game concept to train pilots to keep their heart rates calm during emergencies, since a racing heart can affect decision-making. Researchers are also planning applications in attention management and peak-performance training in aviation.

CyberLearning Technology is working to introduce its product in other health-related sectors where biofeedback training may have benefits. This could possibly include a new type of therapy for aggressive driving behavior, otherwise known as "road rage."

Hearing Is Believing

Twenty-six years ago, Adam Kissiah delivered a medical wonder to the world that has resulted in restored hearing for thousands of individuals, and allowed thousands of others born deaf to perceive sound for the very first time.

Driven by his own hearing problem and three failed corrective surgeries, Kissiah started working in the mid-1970s on what would become known as the cochlear implant, a surgically implantable device that provides hearing sensation to persons with severe-to-profound hearing loss who receive little or no benefit from hearing aids. Uniquely, the cochlear implant concept was not based on theories of medicine, as Kissiah had no medical background whatsoever. Instead, he utilized the technical expertise he learned while working as an electronics instrumentation engineer at NASA's Kennedy Space Center for the basis of his invention. This took place over 3 years, when Kissiah would spend his lunch breaks and evenings in Kennedy's technical library, studying the impact of engineering principles on the inner ear.

Unlike a hearing aid, which just makes sounds louder, the cochlear implant selects speech signal information and then produces a pattern of electrical pulses in a patient's ear. A microphone picks up sounds and transmits them to a speech processor that converts them into digital signals. Although it is impossible to make sounds completely natural, because a mere 22 electrodes are replacing the function of thousands of hair cells in a normal hearing ear, the implant still serves as an excellent rehabilitative mechanism for hearing damage caused by disease, drugs, trauma, or genetic inheritance.

In 1977, NASA helped Kissiah obtain a patent for the cochlear implant. Several years later, he sold the rights of the technology to a company named BIOSTIM, Inc., for commercial development of the innovation. Although BIOSTIM is no longer in business, numerous hearing aid manufacturers have applied Kissiah's patented concept to their cochlear implant products. As the inventor himself puts it, the cochlear implant "only works one way."

It was not until just recently that Kissiah, now retired, started receiving the long-overdue recognition he deserves for this remarkable finding. Kennedy's awards liaison officer, Pam Bookman, who encourages the Center's employees to report their significant contributions, submitted Kissiah for a Space Act Award when she found out his research was drawn from engineering skills honed while with NASA. In 2002, he earned the prestigious award, which included a signed certificate from NASA Administrator Sean O'Keefe and $21,000, the largest monetary award ever given to a single inventor in Kennedy's history.

Perhaps just as rewarding was the opportunity for Kissiah to meet fellow Space Act Award recipient Allan Dianic, an ENSCO, Inc., employee who works with

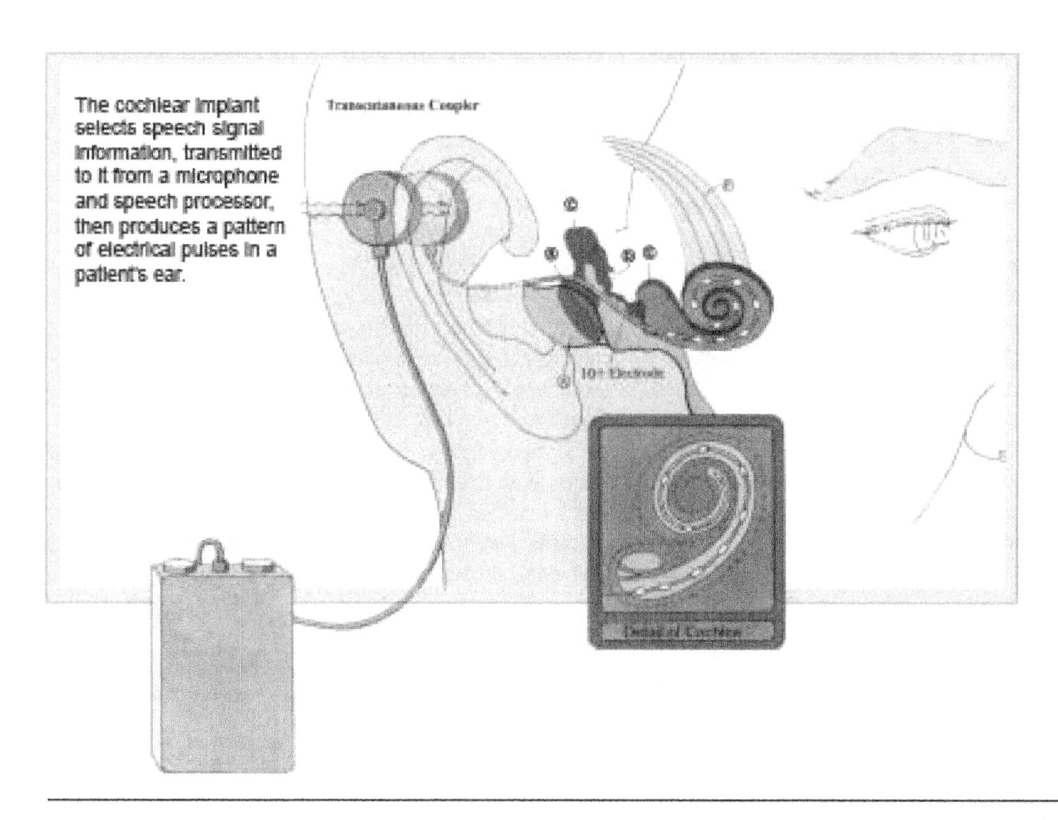

The cochlear implant selects speech signal information, transmitted to it from a microphone and speech processor, then produces a pattern of electrical pulses in a patient's ear.

Adam Kissiah (right), a retired Kennedy Space Center engineer, shows a photo of Allan Dianic's daughter, who has benefited from a cochlear implant that Kissiah developed while at NASA. Dianic (left) is a software engineer with ENSCO, Inc., and a member of NASA's Applied Meteorology Unit. Kissiah received an exceptional category NASA Space Act Award for his technology breakthrough during a technology awards luncheon held at Kennedy's Visitor Complex Debus Center in 2002.

NASA's Applied Meteorology Unit. Dianic's 2-year-old daughter, Victoria, regained full hearing after receiving a cochlear implant just 2 months before the ceremony. It was discovered that Victoria was deaf when she was 9 months old. Now, she can hear for the first time.

In April of 2003, Kissiah was officially inducted into the Space Foundation's U.S. Space Technology Hall of Fame for his invention. Administrator O'Keefe, former astronaut Donald McMonagle, and former astronaut and NASA Administrator Vice Admiral Richard Truly were on hand for the activities; Kennedy Director Roy Bridges, a former astronaut and retired Air Force Major General, accepted the award on behalf of Kissiah and Kennedy (Bridges has since been named director of Langley Research Center). In the face of the latest attention surrounding his technological development, Kissiah has remained extremely humble about his role. "Regardless of what level of participation I had, it is nice to know I contributed to making many lives better," he noted.

The Cochlear Implant Association estimates over 66,000 patients have received an implant, creating what is today a $1.65 billion industry. The American Speech-Language-Hearing Association further states that cochlear implantation consistently ranks among the most cost-effective medical procedures ever reported. In early 2002, popular radio talk show personality Rush Limbaugh revealed during a live broadcast that the "medical marvel" allowed him to hear his show again for the first time since learning he was suffering from near total deafness several months earlier. Limbaugh told his nearly 20 million listeners in October of 2001 that an autoimmune inner-ear disease caused him to lose 100-percent hearing in his left ear and 80-percent hearing in his right ear. Despite this damage, Limbaugh continued his daily broadcasts, responding to callers with the aid of a teleprompter and assistance from his staff.

Heather Whitestone McCallum, who became the first deaf woman to be crowned Miss America in 1995, received a cochlear implant in 2002 and is hearing sounds she never heard before, including the voices of her children. Amy Ecklund, a longtime actress on the daytime soap opera "Guiding Light" and deaf since the age of 6, also regained her hearing with assistance from a cochlear implant in 1999.

Sprung from the mind of an engineer, this medical miracle is a perfect example of how NASA knowledge is boundless and can touch the lives of many in ways unimaginable.

Kissiah's induction into the Space Foundation's U.S. Space Technology Hall of Fame. Left to right: former astronaut Donald McMonagle, Kissiah, former astronaut and NASA Administrator Vice Admiral Richard Truly, and Space Foundation President and CEO Elliot Pulham.

Image courtesy of the Space Foundation. Photographer: Ernie Ferguson

A Robot to Help Make the Rounds

What employee never takes a vacation or a break, never calls in sick, works around the clock 365 days a year, has more than 3 million hours of experience, and is qualified to work in dietary services, radiology and medical record departments, pharmacies, central supply, and laboratories? The answer is a competent, cost-effective robotic courier that enables hospitals to redirect staff to more valuable roles.

The HelpMate trackless robotic hospital courier was designed by Transitions Research Corporation, a think-tank company formed in 1984. Over the course of 10 years, Transitions Research Corporation received funding for this developmental effort from NASA, through seven **Small Business Innovation Research (SBIR)** awards with Johnson Space Center. Notably, Johnson granted the company Phase I and Phase II SBIR funding in 1995 to support the conception of the two-armed, mobile, sensate research-robot, projected to allow development and demonstration of robotic support tasks for NASA on-orbit mechanisms.

In 1997, Transitions Research Corporation went public, and changed its name to HelpMate Robotics, Inc., to emphasize the one product by the same name that was to be introduced to the commercial marketplace. Two years later, Pyxis Corporation, the San Diego, California-based Automation and Information Services division of Cardinal Health, Inc., purchased the assets of HelpMate Robotics, Inc., subsequently acquiring the rights to the HelpMate robotic courier technology.

Known today as the Pyxis HelpMate® SecurePak (SP), the 4-foot-tall, 600-pound trackless robotic courier has evolved significantly from the original design blueprint. According to Cardinal Health, the Pyxis HelpMate SP is the first system of its kind to navigate autonomously through hospitals and other medical facilities—including independently calling and using elevators—without the use of external guidance systems such as fixed tracks or guide wires. It offers an extremely reliable and less expensive replacement for human couriers, while allowing nursing and pharmacy staff and other skilled health care workers to spend less time running errands and more time providing patient care.

The Pyxis HelpMate SP is battery-operated and employs state-of-the-art technology, wireless radio, and proprietary software to guide it from point to point. With a 200-pound payload and various lockable compartments for storage, the robotic courier is capable of smoothly transporting pharmaceuticals, laboratory specimens, equipment and supplies, meals, medical records, and radiology films back and forth between support departments and nursing floors. In a laboratory setting, it delivers test results back to clinicians in a timely manner,

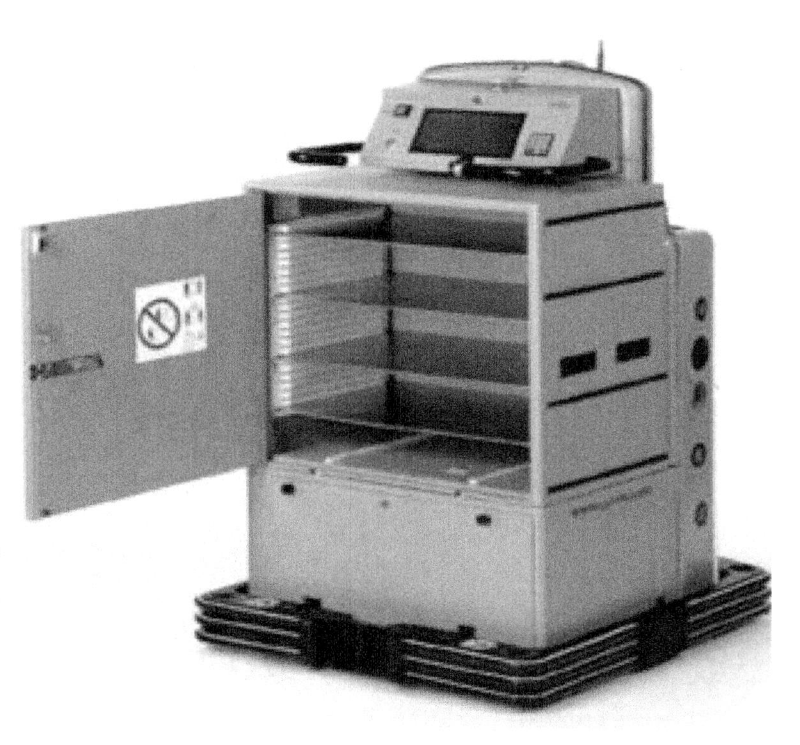

The Pyxis HelpMate® SecurePak robotic courier navigates autonomously throughout medical facilities, transporting pharmaceuticals, laboratory specimens, equipment, supplies, meals, medical records, and radiology films between support departments and nursing floors.

leading to more accurate diagnoses and treatments. Its proven track record for seamless deliveries also helps to eliminate concerns regarding biohazard spills. In the pharmacy, the courier reduces delivery costs, increases productivity, and provides tighter security for medications and supplies.

Pyxis HelpMate SP's easy-to-use color touch screen entitles its human operators to send it to virtually any location in a hospital. Facilities using multiple couriers can centrally manage the robots through a monitoring system to determine location, status, and project destination arrival times. Because they are equipped with radio antennas, the couriers are capable of communicating with each other directly or via a radio frequency Ethernet system.

The source of vision for the original HelpMate was a structured-light vision system, comprised of a camera that sought out objects up to approximately 8 feet in front of the robot. The camera differentiated between obstacles at different distances and altered the course of the robot to maneuver around them. Two infrared strobe lights, positioned 6 inches and 18 inches off of the floor, provided planes of lights for the camera to detect the objects in front of it. A laser scanner replaced the structured-light system in Pyxis HelpMate SP. The new laser scanner provides a wider field of view (180° as opposed to 60° with the camera), finer resolution, and improved reliability. Turn signals, emergency-stop buttons, and contact bumpers are also included as added safety mechanisms.

To date, nearly 100 Pyxis HelpMate units have been sold to hospitals within the United States. The technology presents hospitals and other medical organizations with the financial options necessary to cost-effectively align fiscal goals and strategies, especially when faced with increased labor shortages, fluctuating patient census, and the economic challenges that are currently impacting the health care field.

The robotic courier educes delivery costs, increases productivity, and provides tighter security for medications and supplies.

Images courtesy of Cardinal Health, Inc.

Protein Crystal Growth

Proteins are the chemical building blocks from which all human cells, organs, and tissues are made. They also serve as the hormones, enzymes, and antibodies that help the body fight off invading germs. Determining the structure of a protein enables medical researchers to create pharmaceuticals that will either help or prevent a protein from doing its job. Through a process known as structure-based drug design, researchers use the knowledge of a protein's structure to develop new drugs to treat a variety of diseases. The predominate method of determining a protein's structure is by X-ray crystallography, which involves growing protein crystals and exposing them to an X-ray beam to determine their atomic structure.

In order to rapidly and efficiently grow crystals, tools were needed to automatically identify and analyze the growing process of protein crystals. To meet this need, Diversified Scientific, Inc. (DSI), with the support of a **Small Business Innovation Research (SBIR)** contract from NASA's Marshall Space Flight Center, developed CrystalScore,™ the first automated image acquisition, analysis, and archiving system designed specifically for the macromolecular crystal growing community. It offers automated hardware control, image and data archiving, image processing, a searchable database, and surface plotting of experimental data. CrystalScore is currently being used by numerous pharmaceutical companies and academic and nonprofit research centers. DSI, located in Birmingham, Alabama, was awarded the patent "Method for acquiring, storing, and analyzing crystal images" on March 4, 2003.

Another DSI product made possible by Marshall SBIR funding is VaporPro,™ a unique, comprehensive system that allows for the automated control of vapor

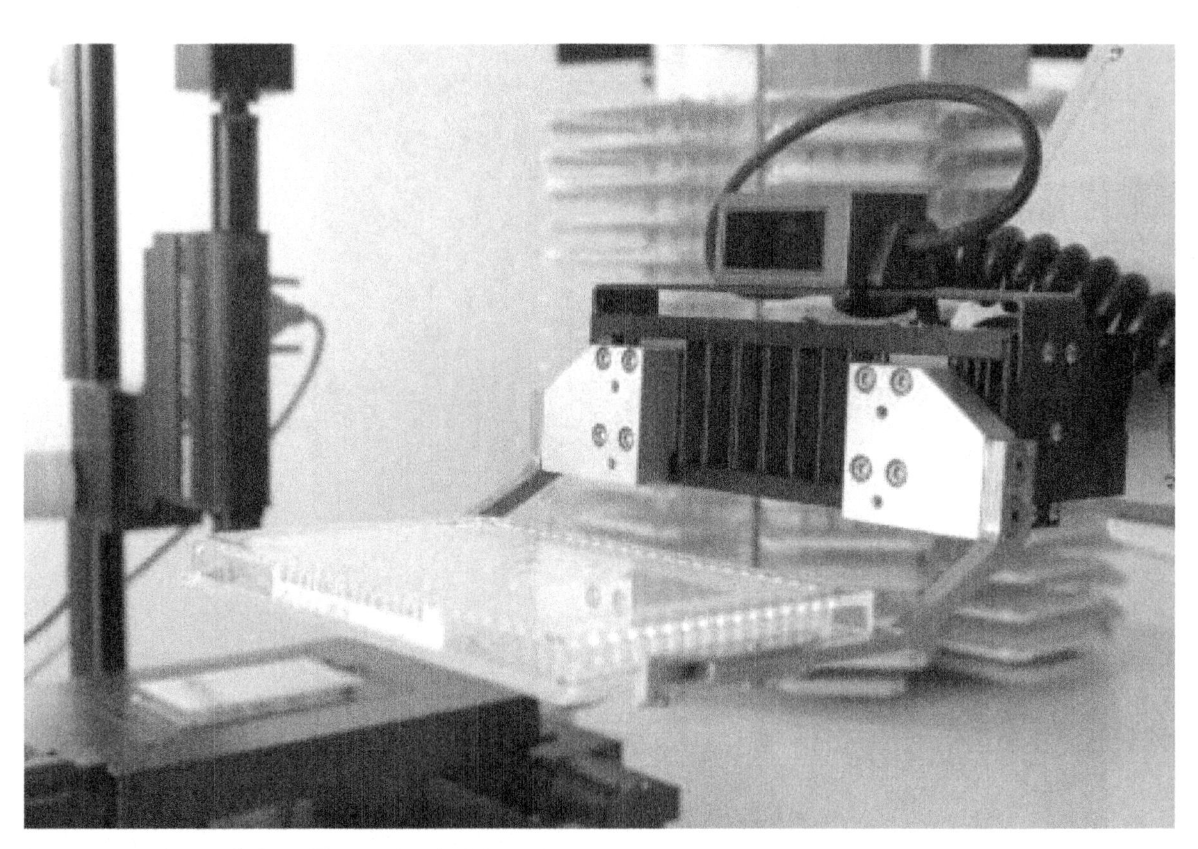

The award-winning CrystalScore™ is an automated image acquisition, analysis, and archiving system for protein crystallization.

VaporPro™ provides solutions for higher quality crystal growth.

diffusion for crystallization experiments. The product contains complete hardware and user-friendly software and was awarded patent protection in June 2002. Its cutting-edge features include individual vapor diffusion profiles for each chamber, as well as automated time-lapse image acquisition, crystal detection, and liquid handling.

With a mission to make drug discovery easier, faster, and more affordable, DSI manufactures and markets products based on the crystallography and structure-based drug design research conducted at the University of Alabama at Birmingham's Center for Biophysical Sciences and Engineering (CBSE), a NASA Commercial Space Center. Formed as a CBSE spinoff company in 1995, DSI has received several SBIR contracts from both NASA and the National Institutes of Health to develop products that support crystal growth for all crystallographic applications, including drug design and protein engineering. The CBSE's commercial research is made possible through NASA's Space Product Development Program, a partnership between NASA, academia, and U.S. industry.

CrystalScore™ and VaporPro™ are trademarks of Diversified Scientific, Inc.

Express Testing Makes for More Effective Vet Visit

From "man's best friend" to the exotic mammals and reptiles that grace the grounds of a zoo, recent improvements in wellness and prevention care are leading to longer and healthier lives for animals, as well as fewer trips to the veterinary office.

The advent of in-office laboratory test systems has pet lovers and animal enthusiasts resting easy, knowing that they are now able to seek medical care and laboratory work, obtain results, and discuss treatment options for their animal companions all in just one 30-minute visit. Historically, veterinary practices would depend on the services of external reference laboratories to process diagnostic test results. With over 60,000 veterinarians representing more than 30,000 veterinary clinics in the United States alone, outsourcing laboratory specimens had a tendency to prolong the turnaround time for results, aggravating clients who are anxiously awaiting feedback from simple blood tests, or causing fear for others who are facing emergency situations with little time to spare.

Already a leading developer, manufacturer, and marketer of point-of-care blood analysis systems for use in human patient-care settings, Abaxis, Inc., of Union City, California, recognized a need for a similar system to address point-of-care diagnostics and other challenges confronting the veterinary market. In crossing over to animal care, Abaxis incorporated elements from its original human blood chemistry analysis system, conceived from NASA research dating back to the late 1970s.

NASA had set out to develop a biochemical analyzer for astronauts to accurately monitor their physiological functions during missions onboard the orbiting Skylab II Space Station. Because of the remote confines of outer space, the analyzer had to be practical for space travel (existing mechanical blood analysis systems were far too large for spacecraft use and would not have functioned properly in microgravity). Something compact yet powerful, reliable, easy to use, transportable, and completely self-contained was the desired end result.

In conjunction with the Tennessee-based Oak Ridge National Laboratory, NASA created an analyzer based on the principals of centrifugal force. The resulting technology successfully achieved separation of human blood cells from plasma, giving NASA the capability to study in-flight samples to gauge astronauts' fluid and electrolyte balance, regulation of calcium metabolism, adrenal function, and carbohydrate, fat, and protein utilization, among other physiological tests. It was patented and exclusively licensed to Abaxis in 1989 for down-to-Earth applications in diagnostics.

Several years later, Abaxis opened the floodgates to new markets and increased revenue opportunities with the introduction of the VetScan® Chemistry Analyzer. According to Abaxis, the VetScan product is the first broad-menu clinical chemistry analyzer designed for point-of-care testing in any treatment setting, including mobile environments, where veterinarians can operate the analyzer from a car-lighter adapter.

VetScan provides veterinarians with the instant diagnostic information they need for rapid treatment decisions, while the patient is still present. Even more, the analyzer completely cuts out the need for follow-up calls and visits when results are not immediately available, and frees up staff for other clinical interventions. In situations where anesthesia and surgery are required, VetScan test results can provide evidence of pre-existing conditions that could possibly lead to anesthetic complications, thereby allowing the veterinarian to optimize conditions prior to performing an invasive procedure and increase the likelihood of a positive outcome.

The VetScan system consists of a 15-pound portable analyzer and pre-configured reagent, or "rotor," disks that enable veterinarians to obtain a clear, comprehensive picture of a patient's condition. The process begins by adding approximately 100 microliters—or about two drops—of whole blood, serum, or plasma to the reagent disk's central chamber (when whole blood is used, pre-analytical centrifugation and other time-consuming steps are eliminated). Considering the small size of many pets,

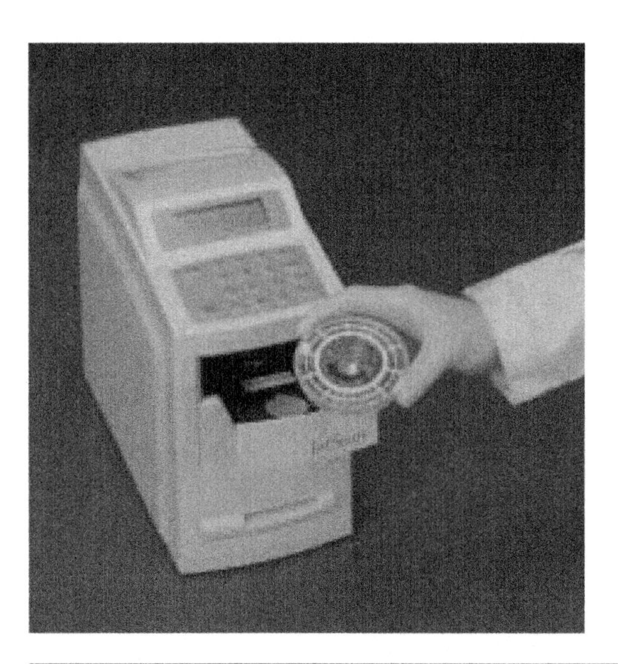

The VetScan® system consists of a 15-pound portable analyzer and pre-configured reagent, or "rotor," disks that enable veterinarians to obtain a clear, comprehensive picture of a patient's condition.

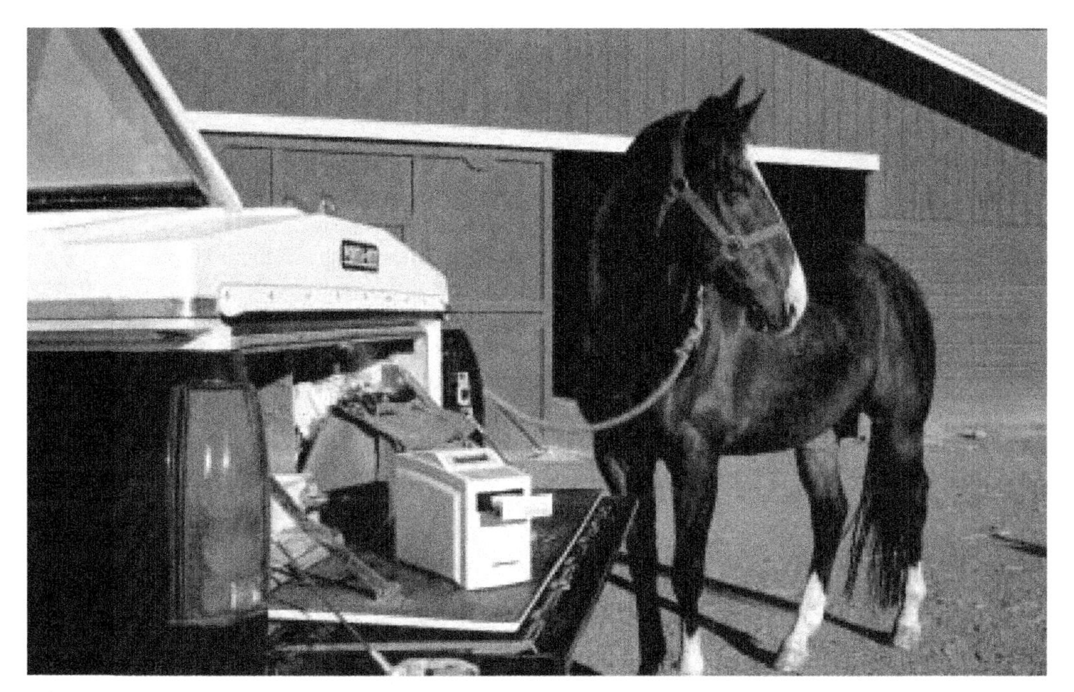

VetScan® is designed for point-of-care testing in any treatment setting, including mobile environments, where veterinarians can operate the analyzer from a car-lighter adapter. A full range of tests is available for almost every species normally treated by veterinarians, including cats, dogs, birds, reptiles, and large animals, such as those in the equine and bovine families.

the tiny sample required is a welcomed change for veterinarians and clients alike. The disk is then ready to be inserted into the reagent drawer, and with less than 2 minutes hands-on time, the VetScan analyzer does the rest.

When the disk is loaded into the analyzer, a series of spinning rotations forces the blood sample into a second chamber where cells separate from plasma. The rotation speed changes, drawing the plasma out into tiny compartments along the perimeter of the disk called cuvettes, which house different chemicals. As the chemicals interact with the plasma, reactions are translated into useful test results for veterinarians.

In less than 15 minutes, a complete profile of results is ready to be interpreted. The analyzer's built-in printer transmits easy-to-read results for immediate review and record-keeping. Accuracy is guaranteed through VetScan's "intelligent Quality Control" (iQC) process, which works nonstop during each run to monitor all aspects of testing, with hundreds of sophisticated automatic checks ranging from basic to complex.

Abaxis currently provides a full range of tests for almost every species normally treated by veterinarians, including cats, dogs, birds, reptiles, and large animals, such as those in the equine and bovine families. The launch of the Avian/Reptilian Profile rotor in June of 2002 has truly filled a market need. Veterinarians caring for birds and exotic animals struggle with the ability to collect an adequate blood sample for traditional biochemistry methods, the company asserts. Medical attention is critical in cases with these animals because they often do not demonstrate symptoms, and therefore, are not seen by a veterinarian until they are seriously ill.

"Every now and then a tool enters the veterinary profession that revolutionizes the way we practice," notes Don J. Harris, a highly revered doctor of veterinary medicine at the Avian & Exotic Animal Medical Center in Miami, Florida. "The Abaxis VetScan is such a tool, especially when dealing with small patients and critical situations, as we often do in exotic animal practice. This device, more than any other in a very long time, has significantly elevated our diagnostic ability."

The VetScan brand and Abaxis' growth in the veterinary market are largely accountable for the company's 400-percent revenue spike in only 6 years, from just over $7 million in 1997 to more than $35 million in fiscal year 2003. To date, the technology has been selected for use by nearly 4,000 veterinary hospitals offering in-clinic testing, and is distributed internationally through a European office and various arrangements in Europe, Asia, and Latin America.

VetScan® is a registered trademark of Abaxis, Inc.

Improving Vision

Many people are familiar with the popular science fiction series Star Trek: The Next Generation, a show featuring a blind character named Geordi La Forge, whose visor-like glasses enable him to see. What many people do not know is that a product very similar to Geordi's glasses is available to assist people with vision conditions, and a NASA engineer's expertise contributed to its development.

The JORDY™ (Joint Optical Reflective Display) device, designed and manufactured by a privately-held medical device company known as Enhanced Vision, enables people with low vision to read, write, and watch television. Low vision, which includes macular degeneration, diabetic retinopathy, and glaucoma, describes eyesight that is 20/70 or worse, and cannot be fully corrected with conventional glasses.

Unlike someone who is blind, a person with low vision retains a small part of his or her useful sight. JORDY enables people to see using their remaining sight by magnifying objects up to 50 times and allowing them to change contrast, brightness, and display modes, depending on what works best for their low vision condition. With this device, people can see objects at any range, from up close to distant. It also provides the flexibility for the user to enjoy theatre, sporting events, and more. JORDY functions as a portable display that is worn like a pair of glasses and as a fully functional desktop video magnifier when placed on a docking stand.

JORDY was inspired by the Low Vision Enhancement System (LVES), a video headset developed through a joint research project between NASA's Stennis Space Center, Johns Hopkins University, and the U.S. Department of Veterans Affairs. Worn like a pair of goggles, LVES contained two eye-level cameras, one with an unmagnified wide-angle view and one with magnification capabilities. The system manipulated the camera images to compensate for a person's low vision limitations. Although the technology was licensed and marketed by Visionics Corporation, LVES was only commercially available for a short time.

In an effort to bring a new and improved low vision headset to the market, Enhanced Vision, of Huntington Beach, California, pursued the development of JORDY. With advances in smaller camera technology, the company significantly increased the head-worn video magnifier's usability, effectiveness, and overall portability. Paul Mogan, an engineer at NASA's Kennedy Space Center, has enthusiastically helped Enhanced Visions' continuing efforts to improve JORDY by contributing ideas and evaluating prototypes.

Legally blind since age 19 due to macular degeneration, Mogan began using the JORDY 1 in 1999. With suggestions for improving the product, he began corresponding with Enhanced Vision. For example, because slight head movements while wearing JORDY would cause the image to jump, Mogan recommended adding

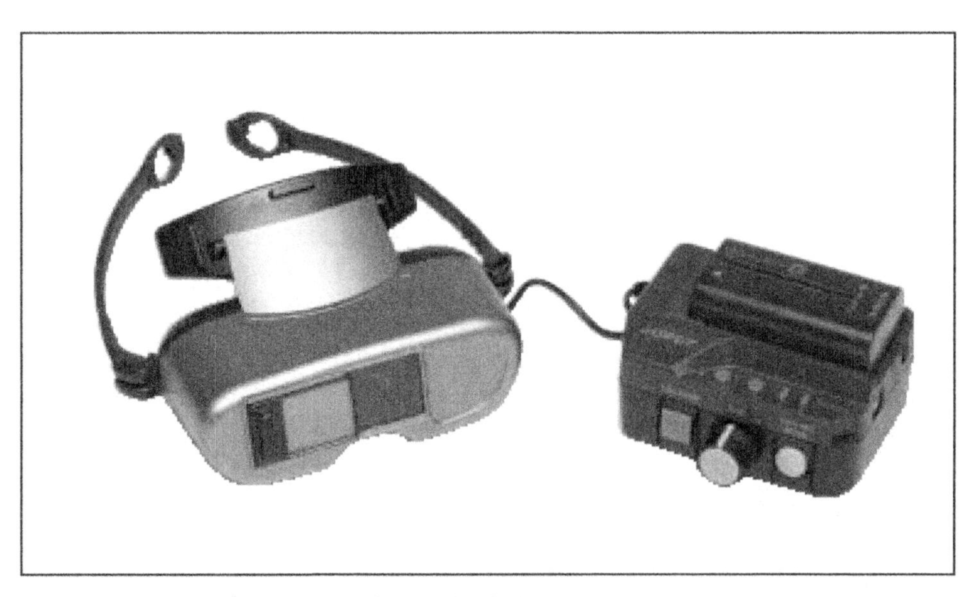

The JORDY™ headset, when worn like a pair of glasses, enables people with low vision to see objects at any range.

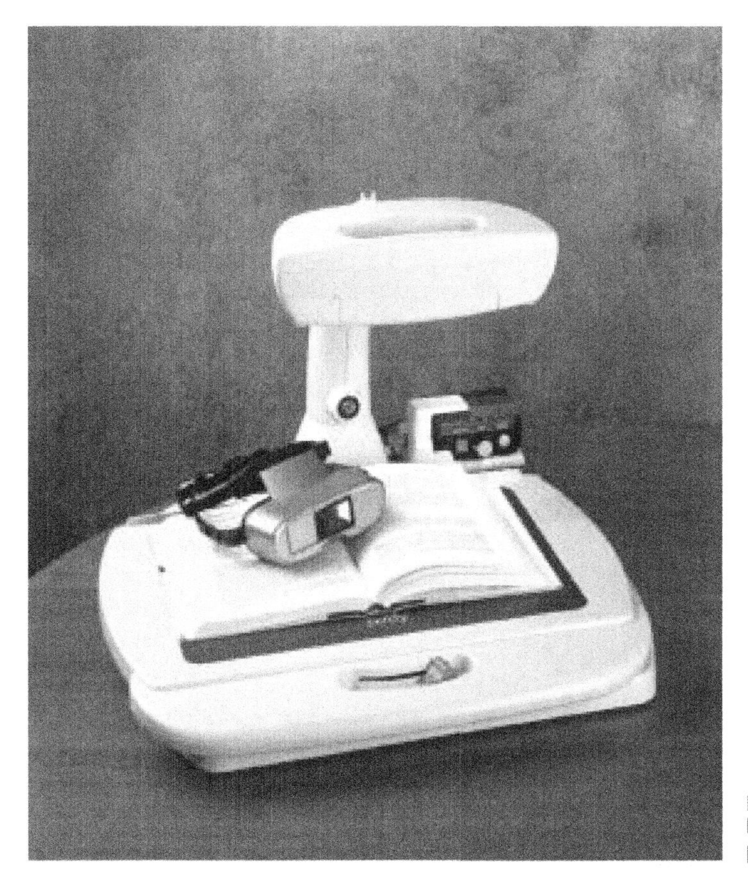

By placing JORDY™ on a docking stand above a book, a person with low vision can read magnified pages that are projected onto a computer screen.

image stabilization to the product. He found the company happy to receive and implement his feedback. When Enhanced Vision developed JORDY 2, a lighter, smaller version of the original device, Mogan again offered suggestions for refinements. One of the enhancements noted for JORDY 3 incorporates an even smaller, high-resolution camera that fits into small, discreet glasses that weigh less than 2 ounces. Increased miniaturization will allow Mogan and others to wear JORDY comfortably for longer periods of time.

JORDY significantly improves the lives of people with low vision by enabling them to pursue their goals. According to Mogan, "As an engineer, I'm always looking for the technologies that are going to give me the edge to keep up. The JORDY has done this more than anything else on the market." He particularly benefits from the product at work whenever he reads for extended periods of time or completes forms. By placing JORDY on a docking stand above a book, Mogan reads magnified pages that are projected onto his computer screen.

Aside from assisting at work, JORDY helps people regain their independence in other ways. The device, when plugged directly into a television set and used in combination with the headset, allows the visually impaired to enjoy television. JORDY also enables people to participate in events, and to see the faces of family and friends. For example, when U.S. speedskater Casey FitzRandolph won a gold medal in the 2002 Winter Olympic Games, his grandfather was able to watch the historic moment from the stands using his JORDY.

An eye care professional or low vision specialist can best determine JORDY's suitability for a patient's individual condition during a routine eye exam. Enhanced Vision works with leading doctors throughout the United States and Canada. Its products are available in over 70 countries worldwide.

Striding Towards Better Physical Therapy

A new rehabilitative device promises to improve physical therapy for patients working to regain the ability to walk after facing traumatic injuries or a degenerative illness. Produced by Enduro Medical Technology, of East Hartford, Connecticut, the Secure Ambulation Module (S.A.M.) creates a stable and secure environment for patients as they stand during ambulation therapy.

S.A.M. is a wheeled walker with a unique harness that supports the patient's body weight and controls the patient's pelvis without restricting hip movement. Electronic linear actuators raise and lower the harness, varying the weight placed on patients' legs. Cable-compliant joints developed at NASA's Goddard Space Flight Center provide S.A.M.'s key element. Consisting of connected cable segments, the joints dynamically connect to the harness, providing stability and shock absorption while allowing for subtle twisting and cushioning.

The late James Kerley, a prominent Goddard Space Flight Center researcher, developed cable-compliant mechanisms in the 1980s for use in sounding rocket assemblies and robotics. This innovative technology uses short segments of cable to connect structural elements. Unlike rigid connections, the cable segments allow movement in six directions and provide energy damping.

Kerley later worked with Goddard's Wayne Eklund and Allen Crane to incorporate the cable-compliant mechanisms into a walker that supported the pelvis. Suffering from severe arthritis himself, Kerley knew that alleviating the weight on the legs was an important part of pain management. The technology allowed the harness to control the pelvis, providing support and stability with compliance that mimicked the movement of the hip joint.

In June 2002, Kenneth Messier, president of Enduro Medical Technology, and Patrick Summers, senior vice president, licensed NASA's cable-compliant technology and walker in order to commercialize the product for medical purposes. The company incorporated the linear actuators into the NASA technology and developed the

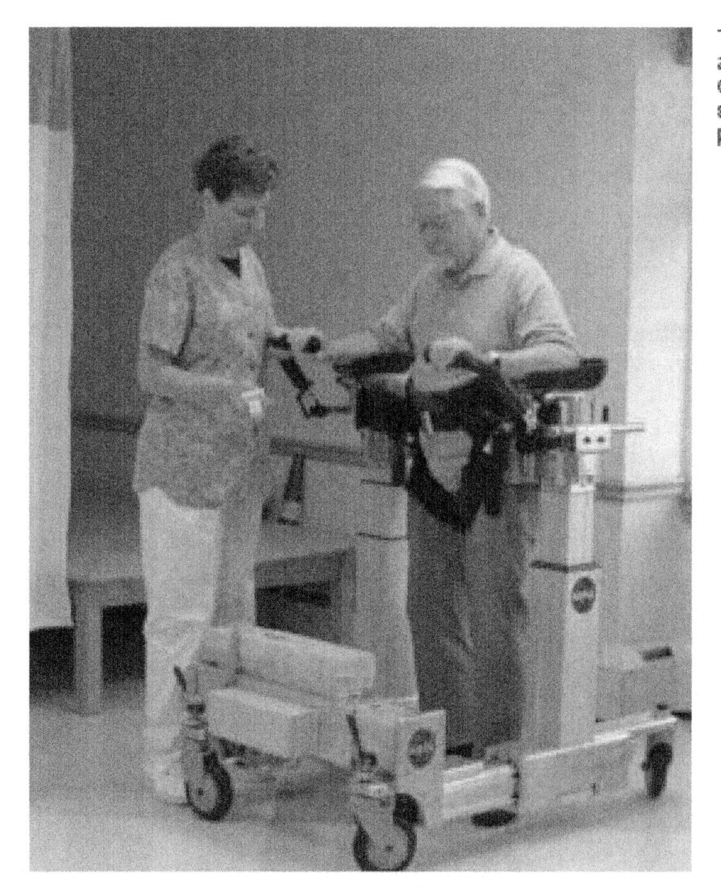

The Secure Ambulation Module creates a stable and secure environment for patients as they stand during ambulation therapy. The device increases staff efficiency, since a single therapist can bring a patient to a standing position.

A unique harness on the wheeled walker supports the patient's body weight and controls the patient's pelvis without restricting hip movement.

adjustable patient harness system, enabling it to introduce S.A.M. to the health care industry in March 2003. The product is marketed towards physical therapists and other health professionals treating patients recovering from traumatic brain injury, stroke, spinal cord injury, and hip or knee replacement. Patients living with severe arthritis, cerebral palsy, multiple sclerosis, Lou Gehrig's disease, and Parkinson's disease can also benefit from S.A.M.

Enduro expects its product to revolutionize physical therapy and restorative nursing. Messier explains, "In the past, patients needing ambulation therapy had to be lifted to standing by one or more physical therapists, and be able to prop themselves up using their arms." As a result, the patients risked falling and their therapists risked back injuries. S.A.M. provides patients with the opportunity to stand and walk in a safe and controlled environment without constant assistance from a therapist. The device reduces patient injuries from falls and increases staff efficiency, since a single therapist can bring a patient to a standing position as well as work with multiple patients at the same time.

While providing a safe environment for gait training, S.A.M. can also help improve a patient's balance, coordination, and endurance. The device may enable patients to have longer therapy sessions and more specialized treatment. Some patients can begin ambulatory rehabilitation sooner since they do not need to prop themselves up with their arms to maintain an upright position. Freeing up the patient's arms also allows the upper extremities to be properly positioned during therapy.

S.A.M. contains several features to make it user friendly. The height and width adjustability accommodates patients weighing up to 500 pounds and ranging from 4 foot 6 inches to 6 foot 4 inches tall. The pelvic harness comes in various sizes and is padded with NASA-developed temper foam for comfort. Attachments for an oxygen bottle, IV pole, and urinary drainage bag are included, as well as an additional upper-trunk harness to provide extra stability for patients with severe balance issues. The electronic linear actuators that adjust the patient's weight bearing can be controlled by the patient or the therapist, and the device includes a digital readout of the adjustments. While patients can use S.A.M. to walk across a room or hallway, it can also be used with a treadmill.

Enduro expects the benefits of S.A.M. to be widespread. According to Messier, the company has shown S.A.M. to hundreds of physical therapists at more than 60 facilities, and all of them indicated interest in utilizing the device. Nona Minnifield Cheeks, chief of the Technology Transfer Program at Goddard, states, "This is a great example of how the research essential for the success of the Nation's Space Program can have clear, tangible benefits in people's daily lives here on Earth."

New Modular Camera No Ordinary Joe

Although dubbed "Little Joe" for its small-format characteristics, a new wavefront sensor camera has proved that it is far from coming up short when paired with high-speed, low-noise applications. SciMeasure Analytical Systems, Inc., a provider of cameras and imaging accessories for use in biomedical research and industrial inspection and quality control, is the eye behind Little Joe's shutter, manufacturing and selling the modular, multi-purpose camera worldwide to advance fields such as astronomy, neurobiology, and cardiology.

In astronomy, Little Joe is used as a wave sensor to eliminate aberrations triggered by wavefront distortions that are known to plague this field with image degradation. Little Joe is also capable of correcting wavefront distortions in medical imaging applications—such as measuring distortions in the human eye—but its high frame rate, high quantum efficiency, and low readnoise properties are really what make the technology an elite member of its camera class. In turn, these properties allow Little Joe to visualize high speed phenomena by optimizing signal-to-noise ratio in light-limited conditions.

Little Joe was not always so little, though. Developed in cooperation with NASA's Jet Propulsion Laboratory under a **Small Business Innovation Research (SBIR)** contract, the wavefront sensor camera underwent radical changes from the time it was merely a concept to the time it was ready to be presented as a commercial charge coupled device (CCD) product. During Phase I of the SBIR contract, Atlanta, Georgia-based SciMeasure worked to adapt a design that would be cost effective, yet powerful in promising efficiency and low-noise operation (the signal generated from CCD cameras contains various noise components that can adversely affect performance). The Phase I camera, however, was essentially a rack of equipment that weighed several hundred pounds, generated roughly 600 watts of heat, and contained components that were imminently obsolete.

SciMeasure and the Jet Propulsion Laboratory were determined to make the camera design as independent as possible from the most critical components, which turned out to be the CCD itself and the analog-to-digital converters that digitize the analog signals from the CCD. Further camera design projects during Phase I led to considerable progress in versatility and modularity

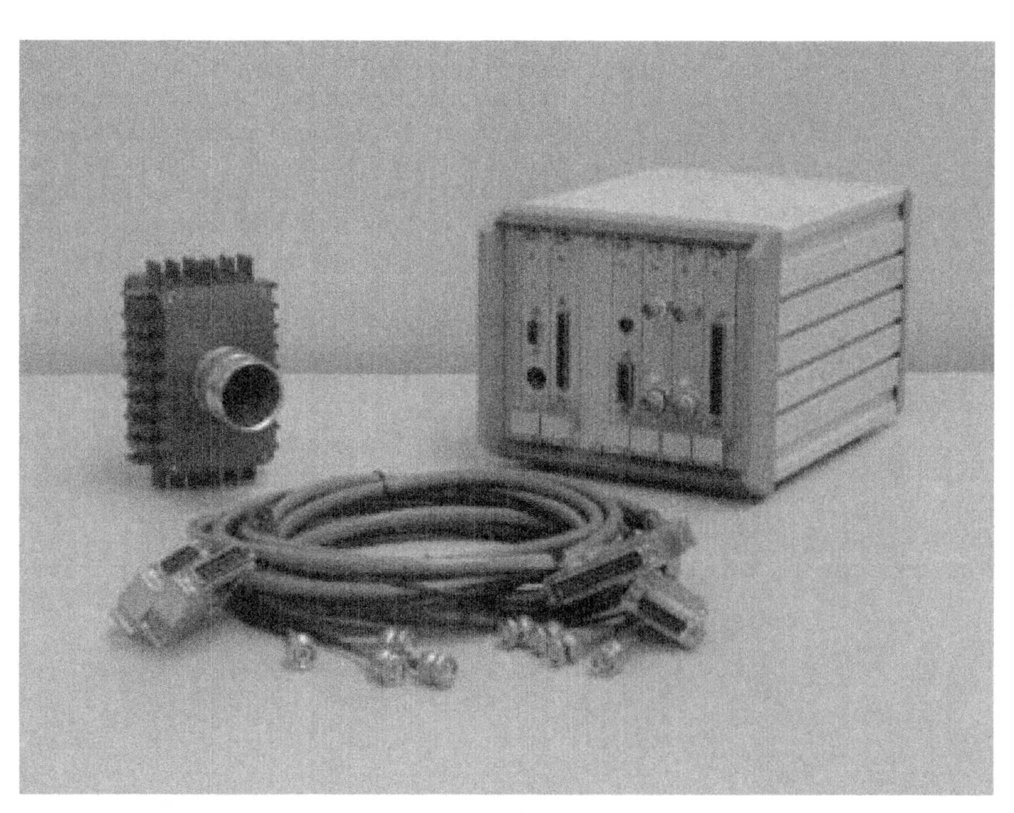

SciMeasure Analytical Systems, Inc.'s "Little Joe" wavefront sensor camera is finding increasing application in astronomy and medicine.

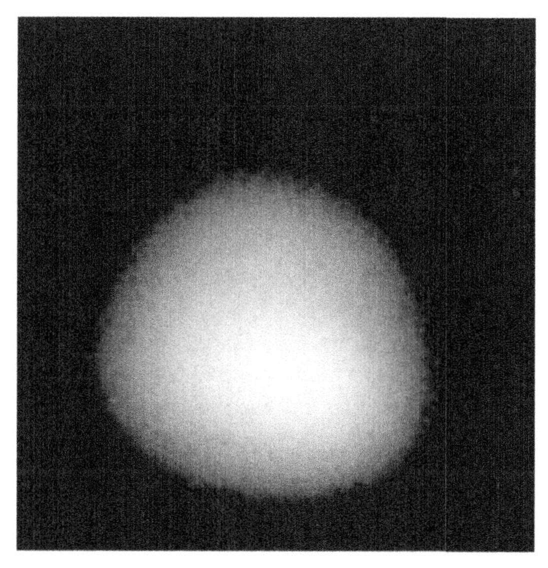

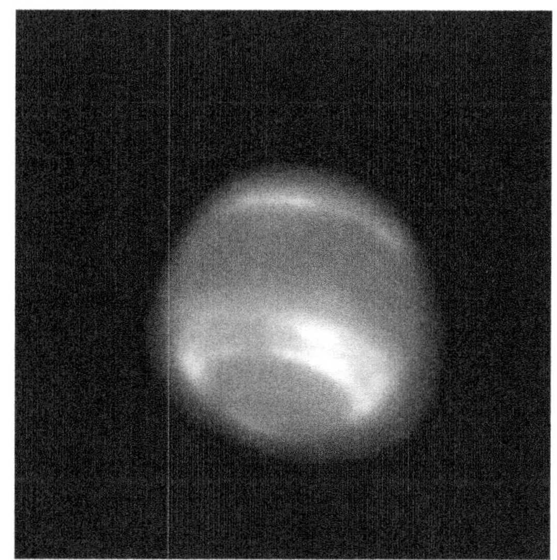

These infrared images of Neptune were obtained by the Jet Propulsion Laboratory/Palomar Observatory Hale Telescope. By using an adaptive optics system that incorporates technology found in "Little Joe," the telescope was able to improve resolution and capture a sharper shot of the planet.

of the technology. The next objective for SciMeasure was to significantly reduce the mass, volume, and power requirements.

The developments that took place in Phase II of the NASA SBIR project translated into a camera that was 95 percent smaller, 92 percent lighter, and used 92.5 percent less power than its first-phase predecessor. Additionally, the camera was configured to run all available scientific CCDs, making it extremely versatile. To address special needs, the camera features an open architecture, allowing end-users to develop replacement or add-in modules.

NASA is using SciMeasure's Little Joe Wavefront Sensor Cameras to support the Jet Propulsion Laboratory/Palomar Observatory Adaptive Optics program, extending the abilities of the 200-inch Hale Telescope located at Palomar Mountain. The cameras have further been selected for the proposed California Extremely Large Telescope (CELT), a joint University of California and California Institute of Technology program aimed at building a 30-meter diameter telescope to generate high-resolution images at short wavelengths. The light-gathering segmented mirror for this terrestrial-based telescope would consist of approximately 1,000 individual mirrors. Future potential application of the cameras exist in the upcoming NASA interferometry explorations, including the 2009 Space Interferometry Mission, which will attempt to determine the positions and distances of stars several hundred times more accurately than any previous program.

South of the stars, the wavefront sensor cameras are finding increasing application at various biomedical and medical research institutions. In late 2001, SciMeasure delivered four commercial Little Joe cameras to RedShirtImaging,™ LLC, of Fairfield, Connecticut, for use in the company's low-light NeuroCCD®-SM neural- and CardioCCD™-SM cardio-imaging systems. Renowned researchers from Yale University all the way to Tokyo University in Japan are utilizing this high-speed, highly sensitive technology to capture the spread of membrane potential and changes in calcium concentration in animal tissue under study. Membrane potential is an important physiological parameter; propagating membrane potential waves is the method that nerve, muscle, and heart cells use to carry information from one end to the other. Calcium concentration is another important parameter because calcium controls many physiological functions, including muscle contraction and communication between nerve cells.

The technology has shown to be imperative for neuroscientists, who commonly perform studies that call for high-speed imaging of fluorescent dyes in the brain at rates of 1,000 to 5,000 frames per second. Cardiovascular scientists can also employ it to monitor abnormal conditions such as tachyarrhythmia. Synchronized operation of two cameras creates an extra functionality for those who would like to simultaneously record cardiac activity using two different dyes or from two different sides of the heart.

With the ability to detect any fast, low-light event with exceptional resolution, Little Joe has demonstrated that it is more than ready to measure up to the many promising commercial applications ahead.

Monitoring Outpatient Care

Each year, health care costs for managing chronically ill patients increase as the life expectancy of Americans continues to grow. To handle this situation, many hospitals, doctors' practices, and home care providers are turning to disease management, a system of coordinated health care interventions and communications, to improve outpatient care. By participating in daily monitoring programs, patients with congestive heart failure, chronic obstructive pulmonary disease, diabetes, and other chronic conditions requiring significant self-care are facing fewer emergency situations and hospitalizations.

Cybernet Medical, a division of Ann Arbor, Michigan-based Cybernet Systems Corporation, is using the latest communications technology to augment the ways health care professionals monitor and assess patients with chronic diseases, while at the same time simplifying the patients' interaction with technology. Cybernet's newest commercial product for this purpose evolved from research funded by NASA, the National Institute of Mental Health, and the Advanced Research Projects Agency. The research focused on the physiological assessment of astronauts and soldiers, human performance evaluation, and human-computer interaction.

NASA's Johnson Space Center granted Cybernet Systems Phase I and Phase II **Small Business Innovation Research (SBIR)** contracts, building upon the company's previous SBIR work on multiple military and Federal Government development projects. The purpose of the NASA project was to enable remote physiological monitoring of space crews. To accomplish this, Cybernet Systems built a miniature portable physiological monitoring device capable of collecting and analyzing a multitude of signals, including electrical brain signals, in real time to monitor astronauts on the International Space Station.

Cybernet's device benefits NASA by immediately correlating the complex interactions between cardiopulmonary, musculoskeletal, and neurovestibular systems in a reduced-gravity environment, leading to a better understanding of the body as a system. In addition, it provides valuable insight into physiological mechanisms, adaptation techniques, and individual responses that occur with exposure to altered gravity environments. This may lead to optimal countermeasure strategies for astronauts to effectively readapt to Earth's environment.

With statistics showing significant improvements in patient outcomes through closer in-home monitoring, Cybernet saw an opportunity to commercialize the physiological measurement and analysis technology. After completing its SBIR work with Johnson in 1998, Cybernet adapted the technology for its MedStar™ Disease Management Data Collection System, an affordable, widely deployable solution for improving in-home-patient chronic disease management. In July 2001, Cybernet Medical announced the general availability of the MedStar interface device and accompanying data collection server, together called the MedStar System.

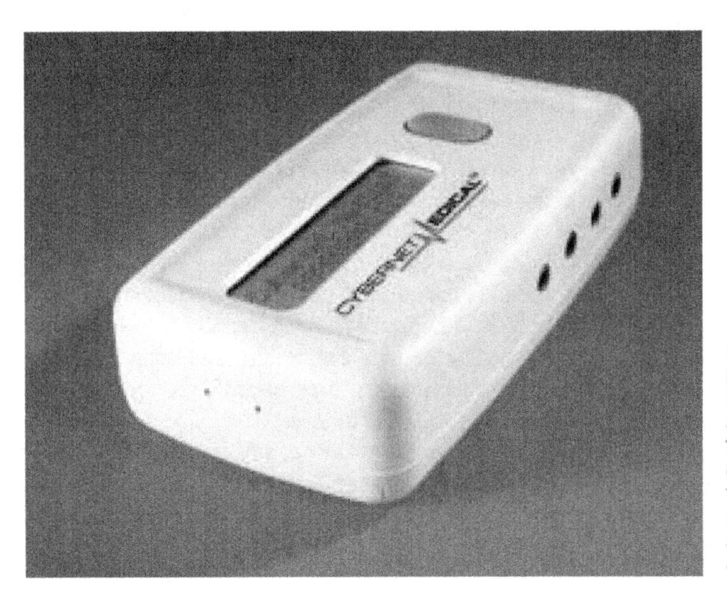

Cybernet Medical's MedStar™ Disease Management Data Collection System is an affordable, widely deployable solution for improving in-home-patient chronic disease management. The system's battery-powered and portable interface device collects physiological data from off-the-shelf instruments.

The battery-powered and portable MedStar interface device collects physiological data from off-the-shelf instruments regularly used at home by chronic-disease patients with high blood pressure, diabetes, congestive heart failure, or respiratory conditions. These devices include weight scales, blood pressure cuffs, and glucose monitors. The MedStar device then securely transmits the data over a standard telephone line to the Cybernet Medical collection server, located at a hospital or a disease management company's facility, for retrieval and analysis. The process enables a health care team to immediately note changes in a patient's condition and make appropriate action recommendations—resulting in fewer patient interventions and emergency hospitalizations.

Measuring 10 square inches and weighing less than a pound, the patient-friendly MedStar device is small and light and operates on standard AA batteries. Since a patient does not need a personal computer or Internet access to transmit MedStar's collected data, the device can be immediately deployed by disease management organizations regardless of patient demographics. MedStar's built-in memory can save several hundred readings, enabling patients on vacation or away from a phone line to continue to take their readings and upload the data when convenient.

Using a database management system, health care professionals can access the data through the Internet in order to remotely manage their patients. Cybernet markets its own data management system, the MedStar Web Server, to retrieve digitized physiological data from a data collection device, such as Cybernet's MedStar Data Collection Server, and uses it to populate a database. It then formats this information for display via a secure Web site, enabling physicians and disease management professionals to analyze changes in a patient's condition. The result is improved patient outcomes and dramatically reduced costs associated with the care of the chronically ill. The MedStar Web Server is available as an addition to the MedStar System, which is also compatible with other commercial database management systems.

MedStar™ is a trademark of Cybernet Systems Corporation.

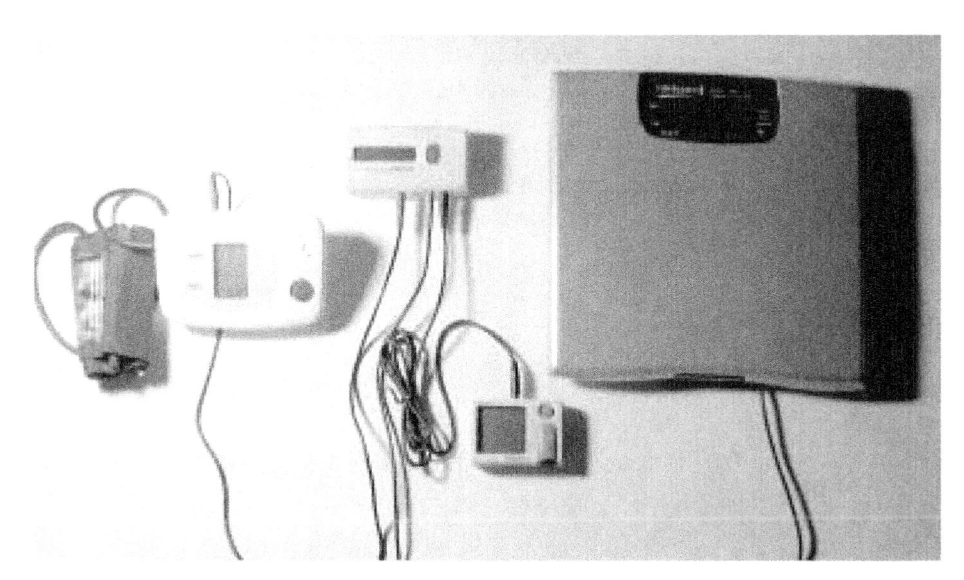

The MedStar™ System consists of an interface device and accompanying data collection server.

InFlight Weather Forecasts at Your Fingertips

A new information system is delivering real-time weather reports to pilots where they need it the most—inside their aircraft cockpits. Codeveloped by NASA and ViGYAN, Inc., the WSI InFlight™ Cockpit Weather System provides a continuous, satellite-based broadcast of weather information to a portable or panel-mounted display inside the cockpit. With complete coverage and content for the continental United States at any altitude, the system is specifically designed for in-flight use.

Hampton, Virginia-based ViGYAN developed the system, originally called the Pilot Weather Advisor, through NASA's **Small Business Innovation Research (SBIR)** program. In the early 1990s, Langley Research Center awarded the company Phases I and II SBIR contracts to develop an innovative concept for a graphical weather advisory system for pilots. Although the Pilot Weather Advisor showed great potential, ViGYAN discovered that the technology was ahead of its time. The system could not become a reality until the cockpit displays and affordable satellite time needed to support it became available.

After investing its own money to keep the project afloat, ViGYAN saw another opportunity to complete the system in 1997 as satellite costs dropped and new cockpit multi-function displays appeared on the market. After more than a decade of work, the company completed its Pilot Weather Advisor in February 2002 through a Phase III SBIR contract with Glenn Research Center's Weather Accident Prevention Project, which is part of NASA's Aviation Safety Program.

In April 2002, ViGYAN sold the Pilot Weather Advisor to WSI Corporation, of Billerica, Massachusetts. According to Keith Hoffler, a former ViGYAN employee who joined WSI as part of the transaction, "We thought about going it alone, but realized that combining our leading-edge technology with the market leader in aviation weather was the smartest way to ensure success." Less than a year later, WSI commercialized the technology as the WSI InFlight Cockpit Weather System.

Gus Martzaklis, Weather Accident Prevention project manager at Glenn, states, "It's gratifying to see NASA-sponsored aviation technologies, like graphical weather displays and satellite data-link communications, come together over the last few years and finally make their way into the marketplace." The WSI InFlight system promises to benefit aviation safety significantly. Martzaklis explains, "Weather contributes to about 30 percent of all aviation accidents. Our research has shown when pilots have real-time, moving weather maps available in the cockpit, they are able to make better, safer decisions faster." With its complete, uninterrupted signal reception, the WSI InFlight system provides a distinct advantage over current ground-based, data-link systems that often have inconsistent signal coverage in large portions of the United States and at various altitudes.

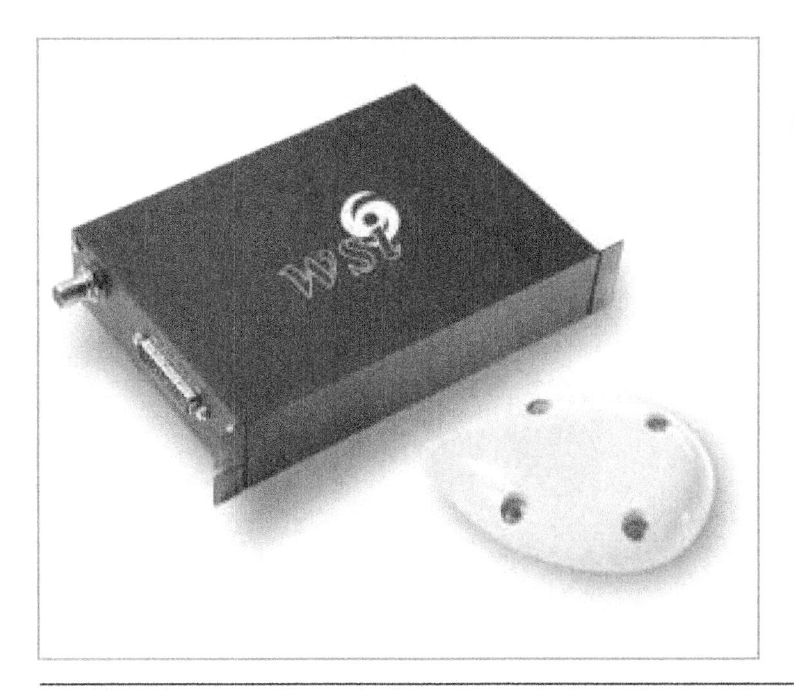

The WSI InFlight™ Cockpit Weather System enables pilots to receive and view high-resolution weather information right inside their aircraft cockpits.

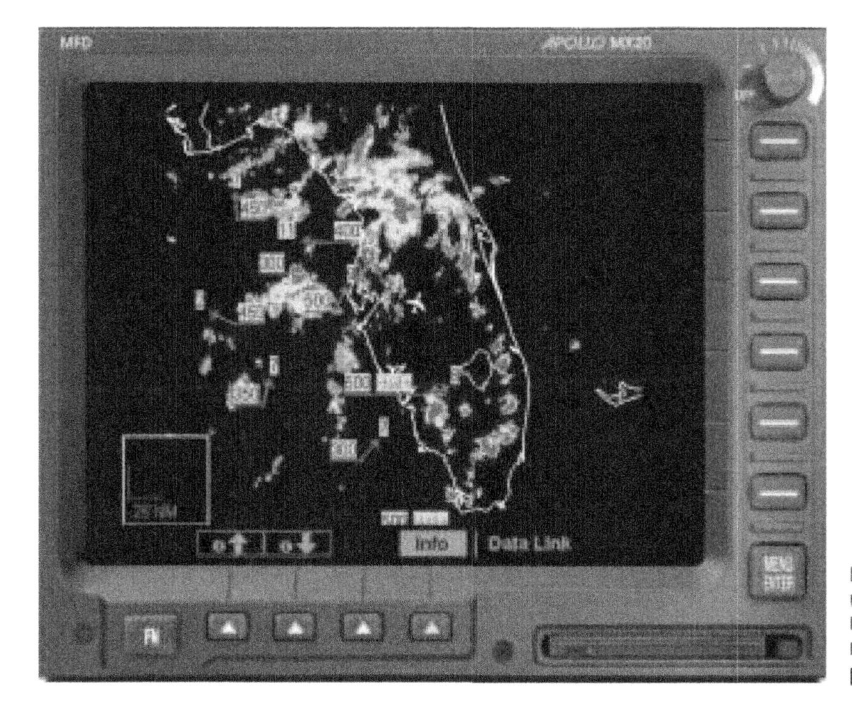

Pilots can view accurate, up-to-date weather information with WSI NOWrad® radar graphics on a variety of panel-mounted, multi-function displays and portable devices.

Pilots using the cockpit weather system receive the most accurate, up-to-date weather information with WSI NOWrad® radar graphics, WSI's flagship national radar mosaic. Updated every 5 minutes, this is the same radar that WSI supplies to its sister companies, The Weather Channel and weather.com, to provide forecasts. Equipped with WSI's special purpose antennas and receivers, pilots can view the high-resolution weather information on a variety of panel-mounted, multi-function displays and portable devices, such as handheld personal digital assistants. WSI's Pocket PC display option allows even the tightest paneled aircraft to utilize the system. After the initial cost of the antenna and receiver, WSI offers flat-rate subscription plans for the service.

WSI InFlight is gaining increased recognition in the aviation community, as several key companies have selected it for their products. UPS Aviation Technologies is working to integrate the system into its MX20 multi-function display for commercial release this year. Rockwell Collins, a global company providing aviation electronics for the world's aircraft manufacturers, selected WSI InFlight to provide weather briefings to aircraft equipped with Collins Pro Line 21 flight deck displays. Northstar Technologies is also integrating the system into its CT-1000 Flight Deck Organizer, which will be marketed to Northstar's growing electronic flight bag customer base.

In addition to improving aviation safety, the same technology in WSI InFlight is forming the foundation for marine and ground transportation applications. WSI is currently developing a boating weather service for mariners that is similar to the cockpit weather service. The company will soon release WSI AtSea,™ which will provide full-color radar, live buoy reports, and offshore forecasts delivered in real time and integrated with a boat's navigation systems. Whether in flight or at sea, WSI's technology keeps people informed about ever-changing weather conditions that can impact their travel and safety.

Putting Safety First in the Sky

Throughout aviation history, a condition known as hypoxia has posed a risk to aircraft pilots, crew members, and passengers flying at high altitudes. Hypoxia occurs when the human body is exposed to high altitudes without protection. Defined as an insufficient supply of oxygen to the body's tissues, hypoxia affects the central nervous system and organs. Brain cells, which are extremely sensitive to oxygen deprivation, can begin to die within 5 minutes after the oxygen supply has been cut off. When hypoxia lasts for longer periods of time, it can cause coma, seizures, and even brain death. Aircraft passengers exposed to either a slow, progressive increase in cabin altitude, or a sudden exposure to high cabin altitude, may show symptoms of inattentiveness, poor judgment, memory loss, and a decrease in motor coordination. Pilots afflicted with hypoxia may not be able to acknowledge the situation or take corrective action, leading to aircraft accidents or crashes.

As a result of technology developed at NASA's Kennedy Space Center, pilots now have a hand-held personal safety device to warn them of potentially dangerous or deteriorating cabin pressure altitude conditions before hypoxia becomes a threat. The Personal Cabin Pressure Altitude Monitor and Warning System monitors cabin pressure to determine when supplemental oxygen should be used according to Federal Aviation Regulations. The device benefits both pressurized and nonpressurized aircraft operations—warning pressurized aircraft when the required safe cabin pressure altitude is compromised, and reminding nonpressurized aircraft when supplemental oxygen is needed.

Jan Zysko, a NASA Applied Research and Development engineer, invented the monitor to give Space Shuttle and International Space Station crew members an additional, independent notification of any depressurization events. Two major incidents—the MIR/Progress collision in 1997 and the Payne Stewart aircraft accident in 1999—reinforced the need for such a device. Zysko, a private pilot himself, also illustrated his invention's necessity in the private sector by citing a

The PCM 1000, a portable, hand-held device approximately the size and weight of a personal pager, alerts pilots to possible hypoxia-causing conditions through simultaneous audio, vibratory, and visual warnings.

The PCM 1000's lighted digital display provides a text message of the warning and the condition causing the alarm.

significant number of hypoxia- and cabin pressure-related incidents contained in accident databases maintained by the National Transportation Safety Board and Federal Aviation Administration (FAA).

As part of the NASA Technology Transfer Program, Kennedy awarded a patent license to Kelly Manufacturing Company, of Grenola, Kansas, to commercialize the monitor. The company previewed the Personal Cabin Pressure Altitude Monitor and Warning System (PCM 1000) at the Experimental Aircraft Association's AirVenture OshKosh 2002 air show after making some modifications and incorporating several new functions. The device was then introduced into the market at the Sun 'n Fun air show in April 2003.

The PCM 1000 is a portable, hand-held, ruggedized device that is approximately the size and weight of a personal pager. It is a potentially life-saving device with simultaneous audio, vibratory, and visual warnings that alert the user to possible hypoxia-causing conditions. In addition, a lighted digital display provides a text message of the warning and the condition causing the alarm. The display also features a low battery indicator. The Altitude Alert Function allows the user to program in a target altitude and a tolerance window to fit the flight situation. This function can also serve as a hypoxia warning system for those who may need an alert at altitudes lower than FAA regulations for oxygen use. While the PCM 1000 is not certified as, nor designed to be, a primary indicator of aircraft altitude, it can serve as a viable alternative for determining altitude in an emergency situation or as a check of instrument function.

According to Zysko, Kennedy's innovation has several other potential commercial uses. Applications beyond the aviation and aerospace industries include scuba diving, skydiving, mountain climbing, meteorology, altitude chambers, and underwater habitats. In the meantime, aircraft pilots can enjoy the extra safety that the PCM 1000 puts right in the palms of their hands.

Smoke Mask

Smoke inhalation injury from the noxious products of fire combustion accounts for as much as 80 percent of fire-related deaths in the United States. Many of these deaths are preventable. Smoke Mask, Inc. (SMI), of Myrtle Beach, South Carolina, is working to decrease these casualties with its line of life safety devices.

The SMI personal escape hood and the Guardian Filtration System provide respiratory protection that enables people to escape from hazardous and unsafe conditions. The breathing filter technology utilized in the products is specifically designed to supply breathable air for 20 minutes. In emergencies, 20 minutes can mean the difference between life and death.

A patent license, acquired from NASA, allowed SMI to utilize a low-temperature oxidation catalyst in its protective breathing filter. The catalyst, developed at NASA's Langley Research Center, converts carbon monoxide to nontoxic carbon dioxide at room temperature, as well as oxidizes formaldehyde fumes, carbon dioxide, and water. Langley's innovation was initially developed for research involving carbon dioxide lasers. In addition to benefiting SMI's escape hood and other air filtration devices, the catalyst promises to have applications in the automobile and aircraft industries and several other areas.

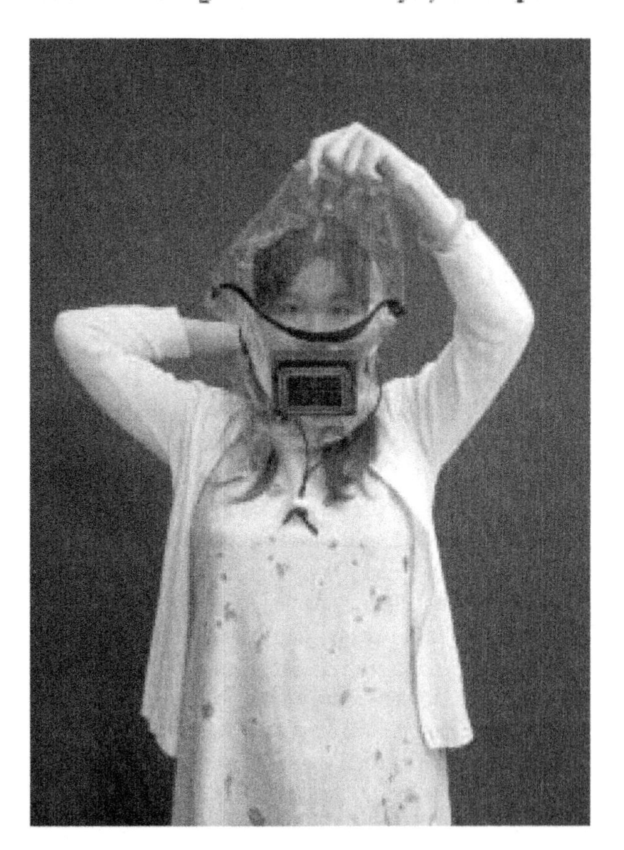

SMI products are designed for emergency use at home, work, and school, as well as for professional firefighting and rescue efforts. The personal escape hood integrates the company's filter with a customized flame-retardant material, and features a clear front to enable vision, an exhalation valve to prevent carbon dioxide build-up, and a drawstring to ensure a tight fit. The hood can be put on in less than 10 seconds, and does not impair hearing or communication. The product comes in both disposable and extended-use models, the latter of which utilizes replacement filters for longer periods of protection.

The Guardian Filtration System is a device designed to fit a wide variety of self-contained breathing apparatuses. While these breathing apparatuses supply the primary air source for firefighters entering smoke-filled or contaminated areas, mechanical failures, air leaks, or "out of air" situations can occur. In these emergencies, the Guardian unit can be quickly attached to the existing face shield. Requiring no additional parts, the unit is easy to apply and uses the same filter technology as the escape hood to provide breathable air for the firefighter.

Prior to the terrorist attacks of September 11, 2001, SMI focused its efforts on devices that would protect users from smoke and carbon monoxide, which have traditionally been the lethal elements in a situation where fire is the primary concern. However, with new threats facing the Nation, SMI is examining how its products can address the concerns of biological, chemical, and nuclear attacks. The company is beginning to add new capabilities to its products, seeking to produce the "ultimate personal escape hood."

The personal escape hood can be put on in less than 10 seconds.

Images Revealing More Than a Thousand Words

A unique sensor developed by ProVision Technologies, a NASA Commercial Space Center housed by the Institute for Technology Development, produces hyperspectral images with cutting-edge applications in food safety, skin health, forensics, and anti-terrorism activities. While hyperspectral imaging technology continues to make advances with ProVision Technologies, it has also been transferred to the commercial sector through a spinoff company, Photon Industries, Inc.

Funded by Marshall Space Flight Center's Space Product Development division, ProVision Technologies originally created its hyperspectral sensor to support human exploration and developments in space. By separating the visible and near-infrared portions of the electromagnetic spectrum, the sensor captures reflected energy from the object it is imaging and splits this energy into more than 1,000 spectral components, or images. The contiguous images can then be analyzed individually or as a set to identify attributes about the object that could not be easily seen otherwise.

For example, these images contain information that may be used to identify a wide range of terrorist weapons, such as toxins and altered passports. ProVision Technologies conducted a pilot study using the sensor to image authentic and forged documents. The technology, which is nondestructive to the papers being analyzed, enabled experts to successfully identify the difference between the inks used for the originals versus the fakes. This advance can help authorities detect counterfeit money, fake passports, and other altered documents being used by terrorists and illegal aliens entering the country.

The technology may also contribute to food safety. Certain molds that grow on grain and corn can create toxins, contaminating food supplies and threatening public health. In order to protect the food supply, new and more rapid methods for detecting these molds are necessary. A ProVision Technologies sensor test captured images of several molds grown by the U.S. Department of Agriculture. Initial results showed that hyperspectral imaging successfully identified molds grown on corn and agar, a gelling agent in food.

After obtaining a patent on its low-cost, portable, lightweight sensor, ProVision Technologies created Photon Industries, located at Stennis Space Center in Mississippi, to commercialize the technology. Photon Industries' relationship to ProVision Technologies gives it access to the hyperspectral technology that is the basis for its product line and services. The company's focus is on developing high-quality, low-cost turnkey hyperspectral imaging systems.

The VNIR 100 is Photon Industries' first product in a series of sensors that will span the spectrum from ultraviolet to thermal infrared to create hyperspectral images. The HyperVisual Image Analyzer,® a graphical user interface-based software package that preprocesses the data collected by the sensor, is included in the system, enabling the end-user to communicate with and control the VNIR 100. This software provides user-friendly tools, including a live preview, an escape function during scanning, geometry control, camera controls, and an image display.

Photon Industries' mission to evolve into being the tool maker of choice for a growing photonics market is aided by the company's consulting and training services. Recognizing that hyperspectral imaging can present challenges for both new and experienced users, the company provides assistance in addressing image acquisition and processing needs. These services aid customers in developing specific applications for hyperspectral imaging.

HyperVisual Image Analyzer® is a registered trademark of ProVision Technologies.

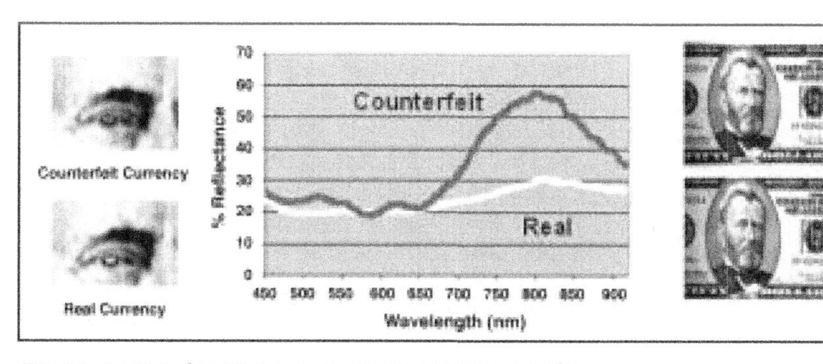

The image shows two 50 dollar bills, one real and one counterfeit. The graph shows the spectral signatures measured from the pupil of President Ulysses S. Grant from both bills.

A Brighter Choice for Safety

Emergency exit signs can be lifesavers, but only if they remain visible when people need them. All too often, power losses or poor visibility can render the signs ineffective. Luna Technologies International, Inc., of Kent, Washington, is shining new light on this safety issue. The company's LUNAplast™ product line illuminates without the need for electricity, maintenance, or a power connection. LUNAplast, which benefited from tests conducted at Johnson Space Center, is so successful that NASA engineers selected it for the emergency exit pathway indicators on the International Space Station (ISS).

LUNAplast™ EXIT signs illuminate without the need for electricity, maintenance, or a power connection.

Available as rigid plastic and acrylic sheeting or flexible vinyl rolls, LUNAplast is an environmentally friendly material available in a full line of screen-printed emergency signs, directional markers, and international safety symbols. It can also be made into strips of various lengths and shapes for illuminating hallways, walkways, and other indoor or outdoor areas. The material is durable, ultraviolet-stable, and fire- and weather-resistant.

Luna Technologies' innovation is the result of advances in photoluminescent (PL) technology. PL materials contain inorganic phosphorescent pigments that absorb almost any kind of light, which is re-emitted when darkness occurs. The effect, called an afterglow, is bright enough to provide a yellowish-green illumination suitable for guiding someone out of a darkened area, such as a stairwell. The nonelectrical and nonradioactive glow is the brightest in the first few hours, but can be seen for days.

While commonly associated with "glow-in-the-dark" toys, PL technology has applications reaching far beyond the children's novelty toys that occasionally utilize it. PL lighting provides emergency exit systems for any setting in which safety is a concern, such as ships, hotels, manufacturing plants, office buildings, tunnels, and mines. As a nonelectric system, PL products provide a backup to standard electric emergency lighting systems.

The zinc sulphide (ZnS) compound used in most standard PL products typically has a limited capacity for storing light energy, causing the illumination to fade within 30 minutes. While many ZnS-based products are utilized in response to mandatory building and fire-safety codes, their limited illumination time curtailed wider voluntary use. LUNAplast changes that, as it emits light over 15 to 25 times brighter than standard PL products, for a significantly longer period of time. ZnS materials have illumination levels of 30 millicandellas to start, before degrading to .032 millicandellas, the lowest light threshold the human eye can perceive, in 3 to 6 hours. LUNAplast, on the other hand, delivers 400 to 600 millicandellas in the initial phase, lasting up to 30 hours before degrading to .032 millicandellas. LUNAplast's initial phase of illumination creates a level of visibility comparable to that produced by a typical electric exit sign.

Luna Technologies achieved this increase in LUNAplast's performance by learning to use strontium aluminate, a nonradioactive metal oxide compound with strong PL characteristics, as the raw material for the technical manufacturing process, in conjunction with the company's proprietary formulation and processing techniques. NASA also played a role in the material's advancement. The NASA Johnson Space Center Lighting Evaluation and Testing Facility tested LUNAplast in its search for a material to create the emergency placards required of the ISS crew member pathway indication system. Luna Technologies took the data from Johnson's tests to improve the product, leading to LUNAplast's selection for the ISS emergency placards.

LUNAplast products were recently installed on the lower walls and floors of the Pentagon as part of the renovation and reconstruction project that took place after the terrorist attacks on September 11, 2001. The supplemental signs mark building evacuation routes to help personnel leave quickly in case conventional electric exit signs lose power or are obscured by smoke, as was the case on September 11 when people faced power loss, darkness, and thick smoke. LUNAplast markings on office floorboards also help guide people who must stay low to the ground to exit. Evacuation maps produced from the glowing material are attached to each door, helping people to follow the escape route. Strips of LUNAplast outline door handles, and closet doors are marked as not being exits.

Luna Technologies produces the LUNAplast 2500 series, its benchmark product, as well as the new 5000 Sun Series. Applications for these products are taking PL technology beyond being a back-up system for electrical emergency exit systems. The LUNAplast EXIT sign meets the National Fire Protection Association's 101-Life Safety Code as a replacement for conventional exit signs, and the Underwriters Laboratory (UL), Inc., an independent, not-for-profit product safety testing and certification organization, lists the product in accordance with its UL 924 standard. This can mean significant savings for companies utilizing the new signs. According to Environmental Protection Agency statistics, installing the LUNAplast EXIT signs instead of electric ones can save a company $20,000 to $30,000 per year, based upon 1,000 electric exit signs. The signs also reduce initial hardware and installation costs, and provide environmental benefits from the reduced power consumption. Combining these savings with the product's life span of at least 25 years truly gives Luna Technologies a reason to glow.

LUNAplast™ is a trademark of Luna Technologies International, Inc.

LUNAplast™ material can be made into strips for illuminating hallways, stairwells, and emergency exits.

A Gold Medal Finish

In February of 1998, the U.S. Speedskating Team was coming off of a performance that yielded a silver medal and a bronze medal at the Winter Olympic Games in Nagano, Japan. Determined to win the coveted gold medal, the skaters looked ahead, setting their sights on a stronger showing at the 2002 Salt Lake City, Utah, games. As fate would have it, the "competitive edge" the team would need to live out these dreams would come compliments of the founder of a company by the same name—in collaboration with NASA.

During the summer of 1999, Darryl Mitchell of Goddard Space Flight Center's Technology Commercialization Office (TCO) met with the U.S. Olympic Committee at the official training facility in Colorado Springs, Colorado, to offer assistance in transferring NASA technologies applicable to Olympic sports. Following the meeting with the Olympic committee, Mitchell was approached by U.S. Speedskating Long Track Program Director Finn Halvorsen, who eagerly voiced his interest in working with NASA to identify a means of improving performance for his team. According to Halvorsen, "If [NASA] can put a man on the moon, surely they can help our skaters."

The U.S. Speedskating organization was looking for new ways to enhance skate technology to ensure that its athletes are the first across the finish line. Areas of possible improvement included reducing the weight of the skate by using new composite materials for the leather boot, adjusting foot positioning for better functionality, and providing better glide and grip for the 1-millimeter-wide skate blade.

Taking these potential advancements into consideration, Mitchell and Halvorsen went to work uncovering NASA technologies that could boost the U.S. team's skating capabilities. Mitchell received a crash course in speedskating, and as a result, generated a lengthy list of promising NASA developments that could benefit the sport. From this list, he and his Goddard TCO partner, Joe Famiglietti, deliberated over whether a NASA mirror-polishing technique could possibly be adapted to the athletes' speedskates. The polishing technique, developed by Jim Lyons, a 16-year optical engineering veteran of Goddard, was derived from the same principles used to create the optics for NASA's science observatories, such as the Hubble Space Telescope (highly polished optics are required by NASA to obtain sharp, clear images in

After bringing home a silver medal and a bronze medal in the 1998 Winter Olympic Games, the U.S. Speedskating organization looked to NASA for new ways to enhance skate technology and ensure that its athletes are the first across the finish line.

space). Lyons, who left Goddard to start up his own optical systems company, PROSystems, Inc., coined his technique "aluminum super polishing." He describes the process as one that allows the polishing of a bare aluminum mirror to create a high-quality aluminum optic that's extremely lightweight and user-friendly.

Lyons and the Goddard TCO team quickly determined that the aluminum super polishing technique was not suitable for speedskate enhancement. This discovery, however, was hardly a setback, as Lyons came through with an alternative polishing process. To gauge the effectiveness of the new procedure, Mitchell and Famiglietti handed Lyons a pair of speedskating blades to see if he could put a shine on the actual edge of each blade. Lyons not only provided the shine, but created a polishing tool used in the blade-sharpening process that improves glide. "For ice skating, the purpose is to reduce the surface roughness, and therefore, you reduce the coefficient of friction, which gives the skate an improved glide," Lyons offers as an explanation for his polishing process.

The new process, which Lyons notes is an extension of how speedskaters typically sharpen their blades, begins by securing a pair of skates in a jig. A sharpening stone is run over the blades, followed by other sharpening tools, each with a progressively finer roughness. During sharpening, a burr begins to form on the blade edge. The burr is removed with a special stone, and the skate blades are ready for polishing.

To prepare the polishing tool, a combination of a polishing compound and a standard lubricating oil is used. When injected through the ports of the tool, the fusion saturates a polyethylene-impregnated felt pad at the bottom, which then comes in contact with the 1-millimeter-thick surface of the blade. The polishing tool is then rubbed quickly and vigorously along the blade, taking just 30 to 60 strokes to improve its edge. Lyons says that the polishing compound, when combined with an abrasive and the pressure applied to the blade, produces a chemical reaction that works to remove surface ridges called "high peaks." He adds that this rough-to-smooth refurbishing procedure is "like comparing a gravel driveway to a finely paved two-lane highway."

The new polishing tool was subjected to a series of glide tests at the Pettit National Ice Center in Milwaukee, Wisconsin, in September of 2001. Speedskates polished with the tool demonstrated an improvement of about 15 percent in unassisted glide over conventionally sharpened skates—like those used in the Nagano Games. Despite the extraordinary results, work was still far from over, and research continued over the next few months.

With the Salt Lake City Winter Olympics just 3 weeks away, Lyons and the PROSystems polishing team finalized a design that met their expectations, and set off to Salt Lake City to work with Halvorsen and the U.S. skaters. Speedskater Chris Witty, the winner of the only two U.S. medals in Nagano, agreed to try the polishing tool. During the trial run, Witty immediately recognized a difference, admitting to an unmistakable increase in speed. On February 17, 2002, she used her newly-polished blades to skate the 1,000-meter race, and glided to victory with a world-record time and a gold medal. In subsequent races, American short- and long-track speedskaters who used the polishing tool also shared a spot with Witty on the winners' podium.

Lyons has since started a second business called Competitive Edge Co., to continue development of the polishing technology for use on skis and in other ice sports like bobsledding, luge, and skeleton. The company is also investigating the possibility of making the jump from the ice rink to the race track with prospective applications in NASCAR and Indy Car Racing.

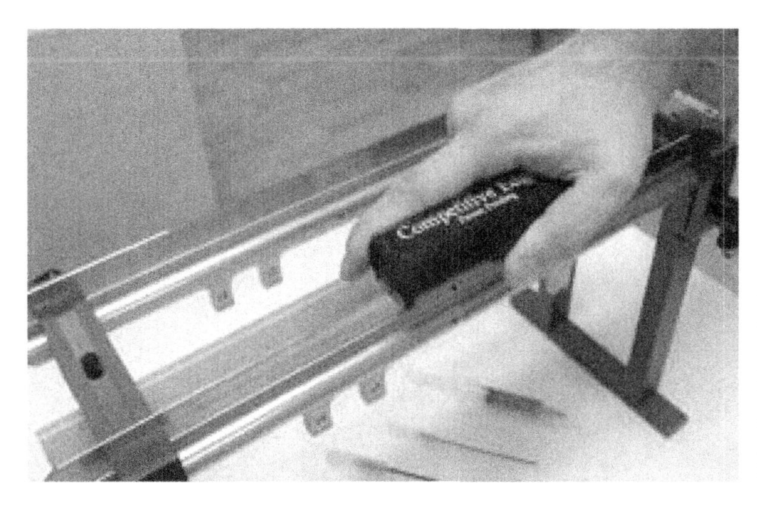

The polishing process, invented by former Goddard Space Flight Center optical engineer Jim Lyons, is an extension of how speedskaters typically sharpen their blades. Following use of a sharpening stone and other sharpening tools, the polishing tool is rubbed quickly and vigorously along a skate blade, taking just 30 to 60 strokes to improve its edge.

Blast-Off on Mission: SPACE

Part of NASA's mission is to inspire the next generation of explorers. NASA often reaches children—the inventors of tomorrow—through teachers, reporters, exhibit designers, and other third-party entities. Therefore, when Walt Disney Imagineering, the creative force behind the planning, design, and construction of Disney parks and resorts around the world, approached NASA with the desire to put realism into its Mission: SPACE project, the Agency was happy to offer its insight.

Mission: SPACE, the newest attraction at Walt Disney World's Epcot theme park in Orlando, Florida, features cutting-edge ride technology that gives guests the incredible sensation of lifting off and traveling through space on a mission to Mars. With input from current and former NASA advisors, astronauts, and scientists, Walt Disney Imagineering developed the attraction to be the first ride system capable of taking the adventurous straight up into simulated flight for a one-of-a-kind astronaut experience.

In 2001, Disney "Imagineers," a name coined to describe the group's unique ability to fuse imagination and engineering, sought advice from Johnson Space Center's Public Affairs Office to anchor the Mission: SPACE attraction with reality-based story elements. The office arranged a tour of Johnson's facility, giving the Imagineers a chance to experience Mission Control and the Advanced Space Suit Laboratory. Teleconferences were conducted between NASA researchers and the Imagineers to discuss the challenges of stepping beyond the lower Earth orbit.

With Mission: SPACE aiming to take guests on a virtual trip to Mars, the Imagineers also sought assistance from NASA's Jet Propulsion Laboratory (JPL) to determine what the Red Planet might look like for landing passengers. JPL provided the Imagineers with satellite imagery of Mars and its terrain. In preparing to design a spacecraft for the attraction, the Imagineers talked with NASA engineers on the future of rocket propulsion technology.

Stepping into the Mission: SPACE courtyard, Planetary Plaza, visitors are transported into the year 2036. The plaza's wall features inspirational quotations from notable figures such as President John F. Kennedy and Columbia Shuttle pioneer Kalpana Chawla. In the attraction's futuristic story line, many countries have joined together to create the International Space Training Center (ISTC), a 45,000-square-foot building featuring a curvilinear exterior that surrounds Planetary Plaza.

In the entrance of the ISTC's Astronaut Recruitment Center, guests see the motto, "We choose to go!", taken from one of President Kennedy's speeches: "We choose to go to the Moon…not because it is easy, but because it is hard." In the Recruitment Center, visitors learn about astronaut training and see a model of the ISTC's X-2 Trainer, the futuristic spacecraft they will board to embark on their mission to Mars. While the X-2 is the creation of Disney Imagineers, it is based on scientific fact and theory provided by scientists, engineers, and future-thinkers from NASA and private industry.

Inside the Space Simulation Lab, a 35-foot-tall gravity wheel slowly turns, containing exercise rooms, offices, work areas, and sleeping cubicles for space teams. Overhead, an authentic Apollo-era Lunar Rover, on loan from the Smithsonian Institution's National Air and Space Museum, is on display as a symbol of mankind's first exploration of another planetary body. A model of the ISTC's X-1 spacecraft (a precursor to the X-2) and a graphic of the X-2 with details explaining the deep space shuttle's functionality also add to the attraction.

After leaving the Space Simulation Lab, aspiring "astronauts" pass through the Training Operations Room, where several large monitors show live video feeds of ongoing ISTC training sessions. They are met by the dispatch officer in Team Dispatch, who assigns them to teams of four people before they are sent into the Ready Room. There, each person is given the role they will assume during the mission—either commander, pilot, navigator, or engineer. The guests are reminded of the importance of training and teamwork before entering

Mission: SPACE at Walt Disney World Resort in Lake Buena Vista, Florida, takes guests on a pulse-racing journey to Mars.

the Pre-Flight Briefing. This area was inspired by the "White Room" at Kennedy Space Center, where astronauts traditionally wait to board their spacecraft.

After a final briefing, each team member enters the X-2 trainer. Mission Control monitors the launch sequence as the capsule moves into launch position, pointed straight up toward the sky. When the countdown reaches zero, the unique and exhilarating ride experience begins. Passengers experience sensations similar to what astronauts feel during liftoff, as they hear the roar of the engines and view computer-generated, photo-realistic imagery based on data taken from NASA's Mars-orbiting satellites. According to Bob Zalk, Walt Disney Imagineer and coproducer of Mission: SPACE, "For the first time, we've combined a unique aerospace technology with classic Disney storytelling, amazing guests with a realistic, one-of-a-kind spaceflight experience." During the ride, the team encounters challenges like those of an astronaut. Each team member must perform the task associated with the role he or she accepted to successfully complete the mission.

After the flight training mission, guests enter the Advanced Training Lab, an interactive play area where they can further test their skills and find out what it takes to be a part of Mission Control. In Mission: SPACE Race, up to 60 people at a time can enroll in a training adventure where two teams, each made up of both astronauts and ground control personnel, race to complete a successful mission. Teams must work together to send their rocket from Mars back to Earth. The area also offers Expedition: Mars, a simulated astronaut obstacle course with a joystick and jet-pack button to help guests explore the surface of the planet. JPL consulted on the game's imagery, creating a realistic Mars landscape.

Story Musgrave, a former NASA astronaut whose career in the Space Program spanned more than 30 years, served as an ongoing consultant on the Mission: SPACE project. According to Musgrave, Disney's new attraction is "a place where guests can imagine our future in space and their role in it, walking in the footsteps of heroes and building on the wealth of technology we've developed to date." Susan Bryan, Walt Disney Imagineer and co-producer of Mission: SPACE, states, "The realism of the experience adds to its uniqueness. Mission: SPACE is very much based in reality; it's a mix of real science and thrill." For NASA, the attraction serves as another source of inspiration for young minds, encouraging them to lead our country, and our world, into tomorrow.

Inside the X-2 capsule, four participants become a team of astronauts working together to fulfill their mission to Mars. During the adventure, which gives participants the sensation of blasting off into space, everyone participates by using joysticks and buttons while viewing outer space through individual video screens.

Visitors become members of Mission Control when they engage in Space Race, a high-energy interactive game that explores the teamwork needed between Mission Control and astronauts in space missions.

Keeping Cool With Solar-Powered Refrigeration

Pioneered by NASA to provide power for satellites and spacecraft, photovoltaics is a viable source of energy used to light over 1 million rural homes around the world. Photovoltaic (PV) cells directly convert sunlight into electricity, without having to utilize limited fossil fuel resources. PV energy contributes to improved air quality and aids in the reduction of greenhouse gases that play a role in global warming. For example, when it displaces coal-fired generation, a common source of electricity among power plants, harmful sulfur dioxide and nitrous oxide emissions are eliminated.

Designed to function in arid to semi-arid regions with at least 5 sun-hours per day, the photovoltaic, direct-drive SunDanzer™ solar refrigerator is a chest-type cabinet with a 105-liter internal volume, a lockable top-opening door, a corrosion-resistant coated steel exterior, and a patented low-frost system.

Most homes running on PV energy, however, employ simplistic lighting systems that are incapable of providing refrigeration. This can be especially troublesome for areas in which no conventional power source exists, including remote automated weather stations, forest stations, and Third World villages.

In the midst of developing battery-free, solar-powered refrigeration and air conditioning systems for habitats in space, David Bergeron, the team leader for NASA's Advanced Refrigerator Technology Team at Johnson Space Center, acknowledged the need for a comparable solar refrigerator that could operate in conjunction with the simple lighting systems already in place on Earth. Bergeron, a 20-year veteran in the aerospace industry, founded the company Solus Refrigeration, Inc., in 1999 to take the patented advanced refrigeration technology he codeveloped with his teammate, Johnson engineer Michael Ewert, to commercial markets. Now known as SunDanze Refrigeration, Inc., Bergeron's company is producing battery-free, PV refrigeration systems under license to NASA, and selling them globally.

Designed to function in arid to semi-arid regions with at least 5 sun-hours per day, the PV direct-drive, or "PV direct," SunDanzer™ solar refrigerator is a chest-type cabinet with a 105-liter (3.7 cubic feet) internal volume, a lockable top-opening door, a corrosion-resistant coated steel exterior, and a patented low-frost system. It uses thermal storage for cooling efficiency, with a direct connection between the vapor compression cooling system and the PV module. This is accomplished by integrating a phase-change material into a well-insulated refrigerator cabinet and developing a microprocessor-based control system that permits the direct connection of a PV module to a variable-speed compressor. The integration allows for peak power-point tracking and the elimination of batteries (thus, the environmental threat of improper battery disposal is eliminated).

For the phase-change material, SunDanzer uses a nontoxic, low-cost, water-based solution with exceptional freezing properties. The variable speed feature allows the compressor to operate longer during the day and better utilize the variable solar resource. A fixedspeed compressor, conversely, can only use about 50 percent of the solar resource, and would not be able to begin cooling as early in the morning or as late in the afternoon, when the sun is low and not shining directly on SunDanzer's solar panel. It would also waste power during solar noon, when the available power is more than the compressor needs to run.

The solar refrigerator's thermal storage material provides 7 days of reserve cold storage, even in tropical climates, or during extensive periods of cloudy weather when sunlight is not available for energy production. Additionally, the unit is manufactured to run on as little as 90 to 120 watts of rated PV power. The combination of a super quiet compressor and fan ensure nearly silent operation, while light indicators on the front of the refrigerator inform users of the status of the thermal reserve. For frequently cloudy regions or areas with less than 5 sun-hours per day, SunDanze Refrigeration offers highly efficient battery-powered refrigerators, as well as freezers. These units run on 12 or 24 volts, direct current, and require a smaller PV or renewable energy system.

Prior to commercialization of the battery-free solar refrigerator in October 2001, NASA installed the original prototype unit in a covered, outdoor location at Johnson in Houston, Texas. For almost 3 years, it was used to store lunches and soft drinks, and was exposed to various experiments, including various defrosting tests. After undergoing several refinements to right the problems encountered, the NASA unit achieved satisfactory results, despite the hot, humid, coastal environment of Houston.

NASA also issued grants to New Mexico State University and Texas Southern University that allowed students and faculty to perform further evaluations of the solar refrigeration technology in a realistic field setting. Testing at New Mexico State included the installation of a battery-free model that consistently cooled up to 6 gallons of drinking water per day, for a period of 15 days, in the dry, desert-like city of Las Cruces. The Texas Southern University tests verified that the solar refrigerator could maintain refrigeration for over a week of continuous overcast weather.

Besides residential homes and stores, applications for the solar refrigerator include cabins, vacation houses, eco-friendly resorts, farms, medical clinics, and street vendor carts. Johnson's Ewert foresees an even wider market with mass production, noting that approximately 2 billion of Earth's inhabitants do not have electricity.

SunDanzer™ is a trademark of SunDanze Refrigeration, Inc.

SunDanzer™ solar-powered refrigerators can keep contents cold for 7 days, even during extensive periods of cloudy weather when sunlight is not available for energy production.

Mapping a Better Vintage

Environmental factors throughout a vineyard can significantly influence the overall quality of wine. Winegrowers have known for centuries that grapes harvested from different areas of a vineyard will produce wines with unique flavors. Affected by subtle differences in the physical characteristics of the vineyard such as microclimate, slope, water-holding capacity, and soil type, even a constant varietal and rootstock of grapes will produce wines with varying color, bouquet, body, and yield, depending upon their location.

Vineyards such as those located in California's Napa Valley tend to be subdivided into relatively large fields or "blocks" that often encompass heterogeneous physical conditions. Since growers typically treat the entire block as a single "minimum management unit" for cultivation and harvest, mapping and monitoring the variability within a block is a concern. Over the last decade, an increasing number of vineyard managers have utilized digital remote sensing and geographic information systems (GIS) to visualize the variability within their blocks. With computer software designed to overlay remotely sensed imagery with environmental and agronomic geographic data on a map, GIS helps growers recognize and explain patterns that might not have been obvious otherwise. GIS can also serve as a valuable archiving mechanism for future reference.

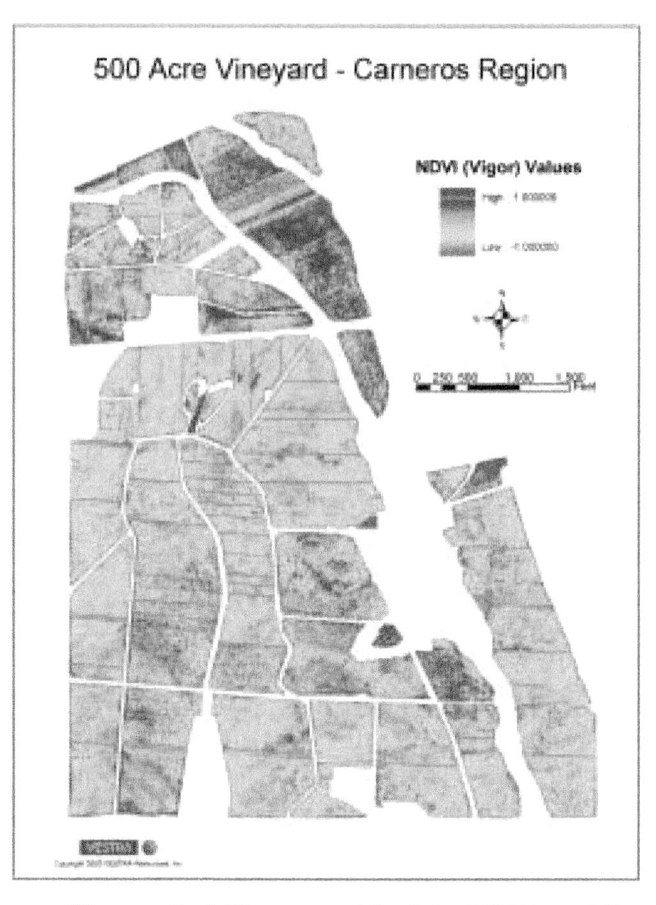

The normalized difference vegetation index (NDVI) is a relative indicator of plant vigor that can help vineyard managers to subdivide a grape harvest for more uniformly mature grapes. In many cases, this improves the quality of the resulting wine lots.

To further develop the use of image technology and GIS for vineyard management support, NASA's Earth Science Enterprise partnered with the U.S. wine and commercial remote sensing industries for a project known as the Viticultural Integration of NASA Technologies for Assessment of the Grapevine Environment (VINTAGE). With project investigators from NASA's Ames Research Center, the California State University at Monterey Bay, and the University of Montana, several prototype products have been developed to support agricultural decisions concerning canopy management and irrigation practice. One key VINTAGE aspect involved the evaluation of satellite and airborne multispectral imagery for delineation of sub-block management zones within a vineyard.

Researchers and vineyard managers analyzed imagery to divide individual vineyard blocks into zones of differing vigor. Using the normalized difference vegetation index (NDVI), a relative indicator of plant canopy density, blocks were subdivided for harvest based upon late-season vigor, resulting in more uniformly mature grapes, and improved quality of resulting wine lots in many cases. NDVI, which is determined by analyzing red and infrared bands from multispectral imagery, has values ranging from -1.0 to +1.0. Higher values indicate more active growth and productivity, while lower values indicate less active vegetation and nonvegetated surfaces. In the vineyard, these index values translate to higher and lower vigor, a factor that frequently relates to fruit characteristics, and ultimately, wine quality. The applied research has shown the feasibility of using such imagery, combined with selective harvest, to move wine lots from lower quality (and value) designations to highest quality reserve programs.

Based on VINTAGE's applied research, VESTRA Resources, Inc., recently released a commercial product known as the Vineyard Block Uniformity Map. Working as a VINTAGE project partner, VESTRA employed the ArcView™ 8.2 and ArcGIS™ Spatial Analyst software from Environmental Systems Research Institute, Inc., to find the percent coefficient of variation (a standard statistical measure) for each block within a 1,000-acre vineyard based on NDVI. The result was a vineyard-level map quantifying block variability, a helpful tool for crop managers.

VESTRA's new map product has already been delivered to several wineries. The maps can serve as an executive summary, allowings managers at companies with large and widespread vineyard holdings to easily identify blocks where new or revised management practices might need to be implemented. Providing a warning, the maps can indicate if a block shows variation over a certain percent. When the Vineyard Block Uniformity Maps are created in consecutive years, a change map can be developed to quantify the increase or decrease in uniformity. They are thus a measurement of success, since managers may use the change maps to determine the effectiveness of mitigation practices. The first Vineyard Block Uniformity Maps were produced for the 2002 growing season, and VESTRA anticipates adding change maps to its 2003 commercial product line. Other prototype products are under evaluation and may be available to growers in the future.

Located in Redding, California, VESTRA is a collaborator on the VINTAGE project along with the Robert Mondavi Winery and the nonprofit Bay Area Shared Information Consortium. All project partners have engaged in extensive industry outreach. VESTRA has worked closely with vineyards and wineries in prestigious U.S. wine regions such as California's Napa Valley and Sonoma County since 1995.

ArcView™ and ArcGIS™ are trademarks of Environmental Systems Research Institute, Inc.

The Vineyard Block Uniformity Map helps winegrowers solve the problem of monitoring variability within a vineyard block.

Digital Images on the Dime

The Earth is ever-changing—above us, around us, and under our feet. Often there are explanations for the shifting behavior of our surroundings, though in some cases, conclusions have not been reached. Many environmental and cultural factors are proven contributors, including climate, natural disaster, erosion, toxic chemicals/pesticides, population growth, land development, sanitation, and urban water runoff. A number of factors may be avoidable to an extent, but at the same time, others are inevitable.

To better understand the changing landscape, scientists, engineers, information brokers, and natural resource specialists depend on aerial and satellite imagery. This large community studies valuable data from digitized images taken from air and space to make important land-management decisions aimed at preserving essential resources and improving quality of life. As crucial as aerial and satellite imaging is to maintaining a healthy global environment, it is an expensive and labor-intensive process that requires compiling and reviewing scores of images to accurately interpret the data.

With NASA on its side, Positive Systems, Inc., of Whitefish, Montana, is veering away from the industry standards defined for producing and processing remotely sensed images. A top developer of imaging products for geographic information system (GIS) and computer-aided design (CAD) applications, Positive Systems is bucking traditional imaging concepts with a cost-effective and time-saving software tool called Digital Images Made Easy (DIME®). Like piecing a jigsaw puzzle together, DIME can integrate a series of raw aerial or satellite snapshots into a single, seamless panoramic image, known as a "mosaic." The "mosaicked" images serve as useful backdrops to GIS maps—which typically consist of line drawings called "vectors"—by allowing users to view a multi-dimensional map that provides substantially more geographic information.

Positive Systems first started working with NASA in 1993 as a partner in Stennis Space Center's Earth Observation Commercial Applications Program, and then again in 1997 through the Center's Scientific Data Buy Program. Both government/industry cooperative programs had specific task orders relating to laboratory testing of the company's Airborne Data Acquisition and Registration (ADAR™) digital aerial photography system for verification and validation of system performance.

In 1997, Positive Systems also received technical assistance on algorithm research for radiometric corrections, specifically to address bi-directional reflectance in aerial photography. This work was conducted under a Space Act Agreement with Stennis, coordinated by the Montana State University TechLink Center, a NASA-funded technology transfer office in Bozeman, Montana. The prototype software developed under the agreement was subjected to rigorous testing and subsequently deemed ready for commercialization under the DIME moniker.

Now marketed worldwide, DIME significantly increases the usefulness of satellite and aerial information by resolving major problems associated with digital imagery. In addition to its mosaicking capabilities, the software reconciles color differences between like features in neighboring images. Variations in colors between images can occur naturally because of lens curvature, solar radiation, and differences in the angle between the sun and the aircraft, which constantly changes as the aircraft flies over the target area and collects photographs. These variations make misinterpretation of imagery features possible. To prevent this, DIME corrects for these internal and external effects and balances the colors of the Earth's features when multiple shots are combined into a single composite image.

Another common problem with digital imagery is that the images are flat, although the subject, the Earth, is round. This can lead to various flaws in pinpointing geographic features. With DIME, users can establish precise longitudinal and latitudinal locations to essentially "pin" the flat images to the round surface of the Earth.

Aerial imagery of the "Boneyard," otherwise known as the Aerospace Maintenance and Recovery Center, Davis-Monthan Air Force Base, Arizona. Twenty-eight raw digital photographs (right) were transformed into a single, georeferenced, color-balanced mosaic (next page) using DIME.®

Image courtesy of Positive Systems, Inc., (www.possys.com).

The technology leverages existing geospatial data sources such as digital orthophotographs, ground control points, and GIS/CAD vector data to achieve lower production costs. By incorporating automated feature-matching algorithms, DIME can quickly assess the necessary corrections to turn raw photographs into orthorectified GIS- and CAD-ready images. According to Positive Systems, recent third party research showed that production with DIME is up to 75 percent faster than traditional methods.

The company is also maintaining low costs with DIME through its unique "pay-as-you-use" software purchasing plan. On top of the pay-per-image benefits, the plan entitles users to buy unlimited copies of the software and receive free upgrades, new releases, and enhanced functionality. The one-time charge per image also gives the consumers unlimited flexibility to generate multiple outputs and edit as necessary.

To promote educational awareness of the Earth's changing environment, Positive Systems has made arrangements with qualified universities and research institutions to provide free image credits for research projects using DIME software. In exchange for the free credits, the universities and institutions work with the company on a quarterly basis to provide written feedback on the research and the results.

DIME is currently employed in over 50 aerial photography, mapping, government, and university research shops, allowing for better natural resource and forestry management, environmental and wetlands monitoring, urban and agricultural planning, and farming. Tobin International, Ltd., of Houston, Texas, purchased DIME and integrated it into a processing method to create low-cost, second-generation orthophotos for use with its proprietary image data management system. Tobin's method provides updated imagery and GIS data for the 7,500-mile Tennessee Gas Pipeline, ensuring safe, efficient operations. The project presents the El Paso Energy Company, also of Houston, with complete visual tools to quickly analyze any encroachments along the pipeline property.

A research team studying infestation patterns of sudden oak death in California used DIME to provide clear indications of dead and dying trees by creating a georeferenced, color-balanced mosaic from 172 separate images. The goal of the ongoing research is to develop a landscape-based risk model of infection that can be tested and exported to other affected areas in the state. In another application involving the Golden State, the data acquisition company, Aerial Information Systems, counted on DIME to help the California Avocado Commission "get to the guacamole." Avocado groves are prone to a root rot fungus that is spread through contaminated soil. DIME aided in the low-cost digital conversion of aerial images by providing a mosaic that helped Aerial Information Systems single out affected and nonbearing trees from a large territory encompassing parts of five counties, ultimately leading to an overall improvement in the classification of avocado inventories. The technology is additionally being used to map illegal immigrant and smuggler trails along the United States-Mexico border.

For future releases of DIME, Positive Systems intends to offer a new pricing structure allowing users to purchase a traditional, one-time-purchase, one-seat, single user copy of the software that does not require image credits for output. The company expects to offer an "Enterprise" multi-user license that does not require purchase of image credits, as well. This new pricing structure will benefit "mega-users" by ultimately reducing their output costs. The current, progressive "pay-as-you-use" plan will stay in place to continue to benefit smaller users.

Expanding on the "Made Easy" aspect of the software, Positive Systems is anticipating an enhanced project set-up through "wizard-like" functionality, allowing users who are unfamiliar with aerial imagery to initiate a project faster and easier, and implementation of Digital Elevation Models and Inertial Measurement Unit data to increase the speed and accuracy of georeferencing capabilities.

Home Insulation With the Stroke of a Brush

Painting the interior or the exterior of a house can be quite an arduous task, but few realize that adding a fresh splash of color to the walls and siding of their homes can lead to reduced energy consumption and substantial savings on utility bills. Hy-Tech Thermal Solutions, LLC, of Melbourne, Florida, is producing a very complex blend of ceramic vacuum-filled refractory products designed to minimize the path of hot air transfer through ceilings, walls, and roofs. The insulating ceramic technology blocks the transfer of heat outward when applied to paint on interior walls and ceilings, and prevents the transfer of heat inward when used to paint exterior walls and roofs, effectively providing year-round comfort in the home.

As a manufacturer and marketer of thermal solutions for residential, commercial, and industrial applications, Hy-Tech Thermal Solutions attributes its success to the high performance insulating ceramic microsphere originally developed from NASA thermal research at Ames Research Center. Shaped like a hollow ball so small that it looks as if it is a single grain of flour to the naked eye (slightly thicker than a human hair), the microsphere is noncombustible and fairly chemical-resistant, and has a wall thickness about 1/10 of the sphere diameter, a compressive strength of about 4,000 pounds per square inch, and a softening point of about 1,800 °C.

Hy-Tech Thermal Solutions improved upon these properties by removing all of the gas inside and creating a vacuum. In effect, a "mini thermos bottle" is produced, acting as a barrier to heat by reflecting it away from the protected surface. When these microspheres are combined with other materials, they enhance the thermal resistance of those materials.

In bulk, the tiny ceramic "beads" have the appearance of a fine talcum powder. Their inert, nontoxic properties allow them to mix easily into any type of paint, coating, adhesive, masonry, or drywall finish. Additionally, their roundness causes them to behave like ball bearings, rolling upon each other, and letting the coatings flow smoothly. When applied like paint to a wall or roof, the microsphere coating shrinks down tight and creates a dense film of the vacuum cells. The resulting ceramic layer improves fire resistance, protects from ultraviolet rays, repels insects such as termites, and shields from the destructive forces of nature.

Hy-Tech Thermal Solutions' proximity to Kennedy Space Center provides the company with the latest advances in the fields of energy, chemistry, and environmental study. Its president, Al Abruzzese, worked for Lockheed Corporation at the Cape Canaveral Air Station in the late 1960s and early 1970s, as a missile team supervisor on nuclear submarines. This position kept Abruzzese in close contact with the activities at Kennedy. Years later, Abruzzese's work as a painting contractor also kept him apprised of the Center's operations. It was during this time that he was exposed to new technologies such as specialized heat-resistant and corrosion-prevention coating systems.

For example, the insulating properties of Space Shuttle tiles immediately came to mind and sparked Abruzzese's interest. He and a coworker asked themselves if it would be possible to incorporate the heat-resistant properties of the ceramic tiles into a commercially available paint product, hence reduce the thermal transfer of treated surfaces. The challenge was many-sided, but Abruzzese discovered that NASA resources, including the use of NASA and university laboratories, were readily available to permit him to conduct extensive research. He soon began to realize the viability of his idea, and pushed forward to overcome the obstacle of finding just the right combination of materials that would provide adequate coverage, would not be too heavy or delicate, could be sprayed, and would not have an adverse reaction with the paint itself.

The insulating additive is available as a stand-alone product that can be mixed into store-bought paints, or as a pre-mixed application in a complete line of factory-blended interior, exterior, waterproofing, and roof coatings.

After a decade of facing storm destruction on Santa Rosa Island, Florida, Mark and Valerie Sigler were awarded a Federal Emergency Management Agency grant to build an energy-efficient, super storm-proof structure, utilizing the latest technological innovations. When construction started on the "dome home" in 2002, the Siglers chose Hy-Tech Thermal Solutions' ceramic additives and coatings to maximize energy efficiency.

By fusing NASA research with his own efforts, Abruzzese selected a variety of ceramics from around the world to intermix and create the "Hy-Tech" Insulating Ceramic Additive he now markets. The Hy-Tech Insulating Ceramic technology is available as a stand-alone product that can be mixed into store-bought paints, or as a pre-mixed application in a complete line of factory-blended interior, exterior, waterproofing, and roof coatings. All products are tested under the harsh conditions of Florida's east coastal region, notorious for its mildew, sulfide gas staining, and hurricane-driven rains.

Hy-Tech Thermal Solution paints and coatings can be used to coat steam pipes and fittings, metal buildings (rust prevention), cold storage facilities (walk-in coolers and freezers), delivery trucks, buses, mobile and modular homes, and RVs and campers. Exterior coatings of the ceramic additive have been applied to trailers housing electronics at Federal aviation locations. The coatings reduce temperature, and thereby lessen the load on the air-conditioning systems inside the trailers. They also provide waterproof surfaces that cut down on moisture in the trailers, keeping the electronic components better protected from the environment. Similar coatings are utilized by the U.S. Forestry Service to insulate and coat underground cables and irrigation systems. These applications are especially effective in preventing rodents and other pests from gnawing through the cables and damaging the underground systems, ultimately averting costly repairs.

Abruzzese encourages homeowners to utilize ceramic-reinforced coatings on their roofing systems to significantly decrease heat inside the attic and home. According to him, "This simple measure can reduce air-conditioning costs significantly. If every home in the United States became just 10 percent more efficient, savings in utility costs would reach into the trillions." He also points out that the Hy-Tech Insulating Ceramics extend the life of a paint coating, and make the painted surface more durable.

As NASA engineers work to improve the Space Shuttle tiles, Abruzzese is watching closely to see how new advancements in this area can continue to influence his own field of work.

Supporting the Growing Needs of the GIS Industry

Remotely sensed imagery collected from orbiting satellites and airborne platforms is playing a vital role in a society driven by a constant need for information. The value of these images rises even more when geospatial features such as buildings, roads, and vegetation are automatically extracted and stored in a geographic information systems (GIS) database to support natural resource, urban, and military planning applications.

Monitoring changes in the Earth's environment from space has long been a primary focus for NASA, but now local, state, and Federal Government agencies, as well as private industry, are increasingly turning to commercial, high-resolution satellite imagery as a source of information to support GIS applications. Nevertheless, a bottleneck exists in this information flow from space that is associated with inflated labor costs and the time required to manually extract geospatial features from digital imagery.

Visual Learning Systems, Inc. (VLS), of Missoula, Montana, has developed a commercial software application called Feature Analyst® to address this logjam. Feature Analyst was conceived under a **Small Business Innovation Research (SBIR)** contract with NASA's Stennis Space Center, and through the Montana State University TechLink Center, an organization funded by NASA and the U.S. Department of Defense to link regional companies with Federal laboratories for joint research and technology transfer. The software provides a paradigm shift to automated feature extraction, as it utilizes spectral, spatial, temporal, and ancillary information to model the feature extraction process; presents the ability to remove clutter; incorporates advanced machine learning techniques to supply unparalleled levels of accuracy; and includes an exceedingly simple interface for feature extraction.

Feature Analyst leverages the natural ability of humans to recognize objects in complex scenes, and does not require the user to explain the human-visual process in an algorithmic form. Since the system does not require programming knowledge, users with little computational knowledge can effectively create automated feature extraction (AFE) models for the tasks under consideration. It offers three levels of automation with its AFE models: the first creates a small training set, explicitly sets up the learning parameters (such as the spatial association settings), and produces an AFE model that is then applied to the remainder of the image; the second creates AFE models that can be shared and then fine-tuned, with a few training examples, for a particular feature extraction program; and the final level of automation involves batch classification, which categorizes imagery with an existing AFE model or set of models. The latter level is considered "full automation," where features are extracted without human interaction.

Other than extraction of single features, Feature Analyst offers many tools for easily creating multi-class extractions, including change detection, three-dimensional feature extraction, data fusion, unsupervised classification, and advanced clean-up and post-processing. With a wealth of options, a user can segment an image into numerous classes, such as water, low- and high-vegetation, and structure.

The U.S. National Imagery and Mapping Agency and the U.S. Forest Service have each completed extensive testing and evaluation of the Feature Analyst with regard to the speed and accuracy of its extraction capabilities. The National Imagery and Mapping Agency concluded that the evaluation of the Feature Analyst software shows substantial benefits for the development of geospatial data from imagery. In one particular assessment, the extraction of land cover and drainage features from commercial satellite imagery was performed approximately five times faster with Feature Analyst than with a standard manual extraction system. While the objective testing concentrated on relatively small scenes, a review of the Feature Analyst's performance over larger regions suggests that the potential time savings in a production setting could be as much as a factor of 100, depending on the homogeneity of the region.

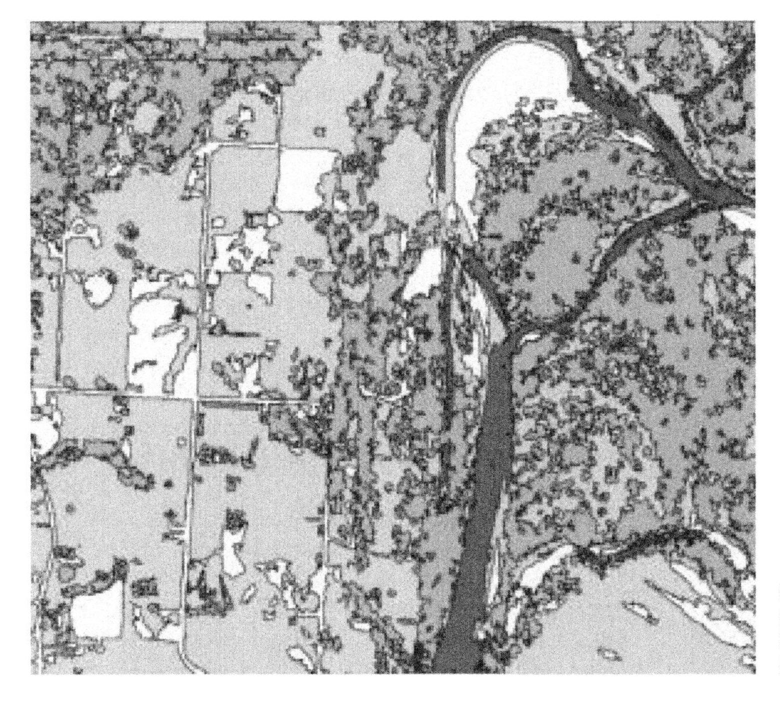

From image to map: Feature Analyst® extracts object-specific, land-cover, and land-use features from satellite and airborne imagery to support the fast-growing GIS industry.

The U.S. Forest Service is currently using the program to map fires and distinguish between burned and unburned foliage, while the U.S. Border Patrol is using it to map trails along national borders. Other areas of utilization include environmental mapping for hazardous waste and oil spill monitoring and cleanup, and transportation planning/asset management for airport runways, wetlands, roads, guard rails, and curbs.

In March of 2003, VLS and Environmental Systems Research Institute, Inc. (ESRI), a global leader in the development of commercial GIS software, signed a strategic agreement that allows ESRI to market and resell Feature Analyst. The pact focuses on solutions for defense and intelligence, homeland security, and environmental, educational, and local government GIS markets.

"Feature Analyst provides a very valuable solution for our users," said Rich Turner, product manager of ESRI's ArcGIS™ software. "Satellite imagery and high-resolution aerial photography are becoming more accessible, not to mention cheaper and more reliable. With a product such as Feature Analyst, our users can better leverage the benefits of imagery as a valuable source of information during the construction and maintenance of their GIS databases." Dr. David Opitz, the chief executive officer of VLS, concurs, adding that the partnership "joins together the market leader in GIS with what is arguably the hottest new product in the remote sensing and GIS industries."

Feature Analyst 3.2 for ArcGIS and another ESRI product, ArcView,™ includes the advanced feature extraction and image classification techniques developed by VLS during the collaboration with NASA, and with additional research from the U.S. Department of Defense.

Feature Analyst is now helping NASA in its critical mission to accelerate and automate the identification and classification of features in digital satellite imagery to support its Earth Science Enterprise mission.

Feature Analyst® is a registered trademark of Visual Learning Systems, Inc.
ArcGIS™ and ArcView™ are trademarks of Environmental Systems Research Institute, Inc.

Putting Fuel Cells to the Test

If research has its way, an electrochemical device capable of converting energy into electricity and heat will become the impetus behind the next generation of automobiles, superseding the internal combustible engine found under the hoods of vehicles that rule the road today.

The thought of fuel cell technology being able to accomplish such a feat may be dismissed as too futuristic by some, but the truth is that fuel cells have been in play as a source of propulsion since the 1960s, when NASA first used them to generate power onboard the Gemini and Apollo spacecraft for extended space missions. Even more unknown is the fact that fuel cells were and continue to be a source of drinking water for astronauts in orbit, since they produce pure water as a by-product.

NASA is recognized for providing fuel cell technology with the initial research and development it required for safe, efficient use within other applications. Fuel cells have garnered a great deal of attention as clean energy converters, free of harmful emissions, since being adopted by the Space Program. Along with automobile manufacturers, universities, national laboratories, and private companies of all sizes have tapped into this technology.

While the primary fuel source for a fuel cell is hydrogen, there are several different types of fuel cells, each having different energy conversion efficiencies. Alkaline Fuel Cells (AFCs), which use a solution of potassium hydroxide in water as their electrolyte, were one of the first classes of fuel cells developed and are still depended upon during Space Shuttle missions. Phosphoric Acid Fuel Cells (PAFCs), considered the most commercially developed fuel cells, are used in hospitals, hotels, and offices, and as the means of propulsion for large vehicles such as buses. Proton Exchange Membrane Fuel Cells (PEMFCs) are similar to PAFCs in that they are acid-based (although the acid is in the form of a proton exchange membrane), but they operate at lower temperatures (about 170 °F, compared to 370 °F) and have a higher power density.

Molten Carbonate Fuel Cells (MCFCs) operate at high temperatures (1,300 °F) to achieve sufficient conductivity. Solid Oxide Fuel Cells (SOFCs), like MCFCs, operate at high temperatures (2,000 °F), but are more ideal for using waste heat to generate steam for space heating, industrial processing, or in a steam turbine to create more electricity. Lastly, Direct Methanol Fuel Cells (DMFCs) use methanol directly as a reducing agent to produce electrical energy, eliminating the need

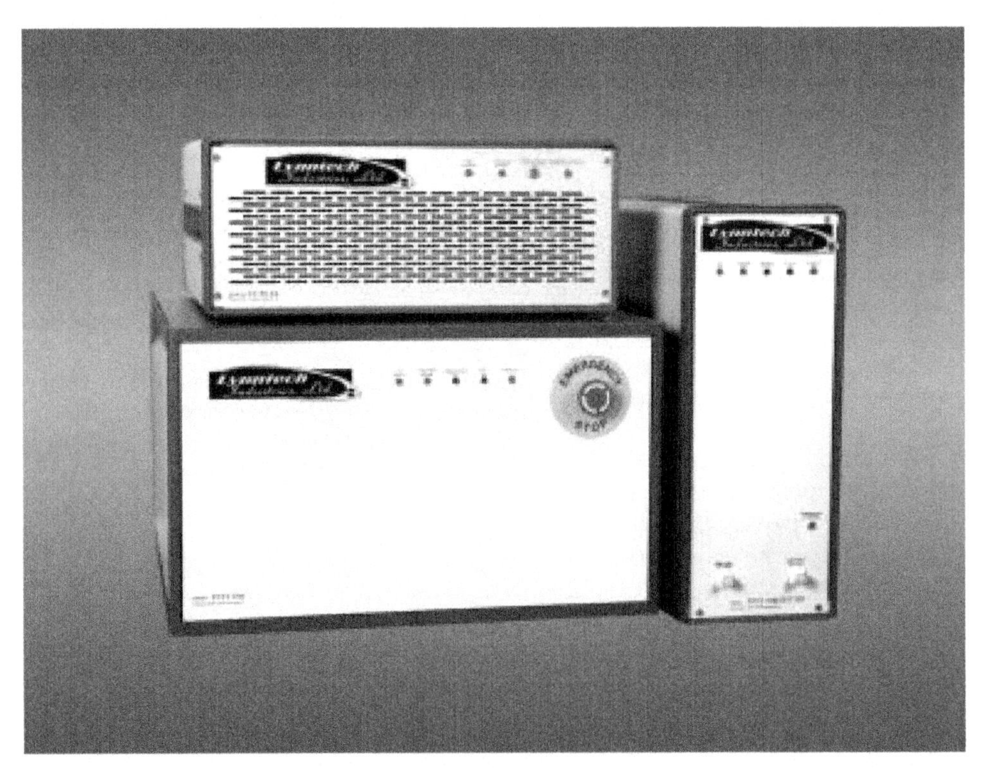

Various modules representing Lynntech, Inc.'s product line of fuel cell test equipment.

for a fuel processor, thus increasing the possibilities for a lighter, less expensive fuel cell engine.

Developers of the various fuel cell technologies require advanced, fully automated, computer-controlled test equipment to determine the performance of fuel cell components, such as electrocatalysts, proton exchange membranes, and bipolar plates, as well as fuel cell stacks and fuel cell power systems. Since 2001, Lynntech Industries, Ltd., an affiliate of College Station, Texas-based Lynntech, Inc., has been manufacturing and selling a complete range of fuel cell test systems worldwide to satisfy customers' demands in this rapidly growing market.

The fuel cell test equipment was invented by Lynntech, Inc., in the early-to-mid 1990s, with funding for design, fabrication, and testing stemming from a Phase II **Small Business Innovation Research (SBIR)** contract with NASA's Glenn Research Center. Glenn awarded the company the SBIR with the intent of utilizing the resulting technology to strengthen NASA's Reusable Launch Vehicle and Space Power programs. First year commercial sales of the fuel cell test equipment were in excess of $750,000, verifying NASA's expense as a sound investment. The test system arising from the work with Glenn has been patented by Lynntech, Inc., and continues to be upgraded to meet current standards.

Lynntech Industries' testing system comes equipped with software called FCPower,™ declared by the company as the most powerful and flexible program for testing in the industry. FCPower enables plug-and-play recognition of hardware, multiple levels of user control, complete automation of configuration and testing, customizable display, and data acquisition and exporting. Even more, the software incorporates safety features that allow for combustible gas monitoring and automatic shutdown of instruments and fuel supply lines.

To match the requirements of individual fuel cell developers, Lynntech Industries adopted a modular approach on designing the test equipment, enabling custom solutions with standard equipment. This entitles customers to select specific modules they may need for any given fuel cell application. Accordingly, Lynntech Industries provides a selection of "all-in-one" test systems and function-specific modules. The components of the company's fuel cell test system include an electronic loadbank; a reactant gas humidifier; gas mixing, handling, and metering systems; instrumentation input/output; methanol and hydrogen test kits; tail gas handling; thermal management; and a cell voltage monitoring buffer board.

It remains uncertain when exactly the average consumer will be able to fully appreciate the impact that fuel cells are making to preserve the environment, but Lynntech, Inc., and Lynntech Industries are in position to bring this moment of realization one step closer to reality.

FCPower™ is a trademark of Lynntech Industries, Ltd.
Flightweight™ is a trademark of Lynntech, Inc.

A variety of fuel cell components, like the Lynntech Flightweight™ Fuel Cell Stack shown here, require top-of-the-line test equipment to determine their performance.

Easy and Accessible Imaging Software

DATASTAR, Inc., of Picayune, Mississippi, has taken NASA's award-winning Earth Resources Laboratory Applications Software (ELAS) program and evolved it into a user-friendly desktop application and Internet service to perform processing, analysis, and manipulation of remotely sensed imagery data.

NASA's Stennis Space Center developed ELAS in the early 1980s to process satellite and airborne sensor imagery data of the Earth's surface into readable and accessible information. Since then, ELAS information has been applied worldwide to determine soil content, rainfall levels, and numerous other variances of topographical information. However, end-users customarily had to depend on scientific or computer experts to provide the results, because the imaging processing system was intricate and labor intensive.

In 1992, Stennis' Commercial Technology Program made ELAS available to DATASTAR under the Freedom of Information Act, which allows federally developed technologies that are not patent protected to be transferred to U.S. companies. The company adapted the NASA technology into the DATASTAR Image Processing Exploitation (DIPEx) program, making the ELAS program simpler and more accessible to general end-users.

DIPEx can separate and provide specifics of imagery data, such as data classifications, false color composites, soils, corridor analysis, subsurface vegetation, data enrichment, mosaics, and geographical information systems (GIS). The program has enhanced mapping capabilities and colorized data for depth. Data generated by DIPEx are compatible with all of the GIS software packages on the market.

DATASTAR offers the DIPEx Service Delivery System, a subscription service available over the Internet, to provide normalized geospatial data in the form of products. Upon opening an account, users can either request a deliverable product from DATASTAR or access the data sets on their own computers. The images or maps that are created through DIPEx are dynamically generated based on the layers and combinations of data chosen. Users simply click a button to add or subtract a layer of information, and create an information product or decision product. The system, structured to allow hundreds of people to access it simultaneously, is on a secure server to protect its intellectual property and the personal data of its subscribers.

DIPEx uniquely incorporates NASA's ELAS into a format usable on most of today's popular systems, from PCs to larger UNIX and LINUX servers. The company also added interface ability to standard file structure and sequel database structure for control. The dimensionality of DIPEx internals assures that the software is current with leading-edge hardware offerings in the computer industry. DIPEx expands the parameters of the original ELAS design, enabling it to address current local and regional database requirements. The product also has the ability to read all of today's high-resolution imagery.

End-users interested in spatial data, such as soil content, rainfall levels, and other variances of topographical information, but who do not have the time or expertise to manipulate the data, will appreciate the convenience of DIPEx. A particularly strong product attribute is the ability to manipulate both raster and vector data. By combining these two types of data, DIPEx performs complex ad hoc queries of specific geographical areas under the control of the investigator.

One of the largest applications of DIPEx data is prescription farming. DIPEx generates data for farm consultants to control field machinery that apply pesticides and water. By offering the service over the Internet, DATASTAR sees the product as a tremendous resource for consultants that work with farmers to maintain the health and yield of crops and land. As subscribers to DIPEx, crop consultants can access the program with specific input parameters and create an information product about a tract of land. The consultant would then be able to make recommendations to the farmer regarding specific soil nutrient additives, irrigation, or pest control.

From analyzing geographical data to determine rainfall levels to providing data that will improve crops, DIPEx presents information for researchers, scientists, and agriculturalists to better understand Earth's valuable resources.

These images are DIPEx-classified products generated from data acquired for a major U.S. timber management company. The objective was to "count" the number of trees in a young timber stand. The DIPEx system can determine the regeneration of a timber stand remotely, and determine the advocacy of replanting a stand. This process makes foresters more productive and accurate in doing their work. The two images were generated using the DIPEx Parallelepiped and Point Cluster classifiers, respectively. Both classifiers agreed that there was a 44 percent stand of trees and recommended that the stands be replanted for optimum return.

Faster Aerodynamic Simulation With Cart3D

A NASA-developed aerodynamic simulation tool is ensuring the safety of future space operations while providing designers and engineers with an automated, highly accurate computer simulation suite. Cart3D, co-winner of NASA's 2002 Software of the Year award, is the result of over 10 years of research and software development conducted by Michael Aftosmis and Dr. John Melton of Ames Research Center and Professor Marsha Berger of the Courant Institute at New York University.

Cart3D offers a revolutionary approach to computational fluid dynamics (CFD), the computer simulation of how fluids and gases flow around an object of a particular design. By fusing technological advancements in diverse fields such as mineralogy, computer graphics, computational geometry, and fluid dynamics, the software provides a new industrial geometry processing and fluid analysis capability with unsurpassed automation and efficiency.

Before the development of Cart3D, grid layouts used to analyze the designs of airplanes and spacecraft needed to be hand-generated, requiring months or even years to produce complex models. Engineers develop these grids to calculate flow fields surrounding vehicles like the Space Shuttle. Cart3D automates grid generation to a remarkable degree, reducing simulation time requirements significantly. The software streamlines the conceptual and preliminary analysis of both new and existing aerospace vehicles. The Cart3D package includes utilities for geometry import, surface modeling and intersection, mesh generation, and flow simulation.

Through a joint agreement with the Ames Commercial Technology Office, ANSYS, Inc., a global innovator of simulation software and technologies designed to optimize product development processes, has integrated the Cart3D product into its ICEM CFD Engineering (an ANSYS subsidiary) product suite for commercial distribution. The package includes several new features, including a graphical user interface for analysis setup. It also incorporates the company's technology for geometry acquisition, repair, and preparation. Computer-aided design (CAD) geometry is directly imported with the company's Direct CAD Interfaces. Designers and engineers can automatically set up and run suites of simulations based on parametric changes to CAD geometry models.

Today, several commercial users, NASA, and leading universities such as the Massachusetts Institute of Technology, Johns Hopkins University, and Stanford University, benefit from Cart3D's capabilities. Northrop Grumman and Raytheon apply Cart3D to the analysis and conceptual design of military vehicles and commercial aircraft. Simulations generated by the program help to identify and fix problems with transport aircraft and helicopters. At Johnson Space Center, Cart3D simulates various crew escape configurations for NASA's Space Launch Initiative program.

ANSYS intends to expand Cart3D's applications well beyond traditional aerospace uses, to aerodynamic and fluid flow simulations in automotive, turbomachinery, electronics, and process industries.

Cart3D automates the grid layouts for aircraft and spacecraft design analysis.

Promising More Information

When NASA needed a real-time, online database system capable of tracking documentation changes in its propulsion test facilities, engineers at Stennis Space Center joined with ECT International, of Brookfield, Wisconsin, to create a solution. Through NASA's Dual-Use Program, ECT developed Exdata, a software program that works within the company's existing Promis•e® software. Exdata not only satisfied NASA's requirements, but also expanded ECT's commercial product line.

Promis•e, ECT's primary product, is an intelligent software program with specialized functions for designing and documenting electrical control systems. An add-on to AutoCAD® software, Promis•e generates control system schematics, panel layouts, bills of material, wire lists, and terminal plans. The drawing functions include symbol libraries, macros, and automatic line breaking. Primary Promis•e customers include manufacturing companies, utilities, and other organizations with complex processes to control.

NASA uses Promis•e to create drawings and schematics at several Stennis test facilities. These facilities test the Space Shuttle main engines, rocket propulsion systems, and related rocket engine components, with each test typically having different measurement and control system requirements. As a result, modifications need to be made to accommodate changing test articles and data requirements. Since the Promis•e software was limited to 120 storage values with every schematic symbol, NASA needed increased values to accurately and efficiently document all of the measurements, control system changes, upgrades, and data associated with each test.

In response to Stennis' need, ECT developed Exdata, an external database program that expands storage capability, automates the design process, and reduces turnaround time for test requirements. The program links a Promis•e schematic symbol with a second Microsoft® Access database file. This second external file greatly increases the amount of information available to a user, and allows direct access for adding additional data, structure, and programming. Changes to the data can be made either in the database or on the drawing.

After collaborating with Stennis, ECT now sells Exdata as a part of its product family. Exdata's main benefit allows customization of data storage and display. By creating custom forms in Access, users can manipulate and display the information most important to them. For example, Exdata keeps maintenance information on a device to determine hours of usage and when the next scheduled maintenance is required.

The level of customization that someone can achieve with Exdata is only limited by the functionality in Access and the person's own ability to apply that function. With this type of control, ECT's Promis•e and Exdata software products are leading the way for faster, more efficient design solutions.

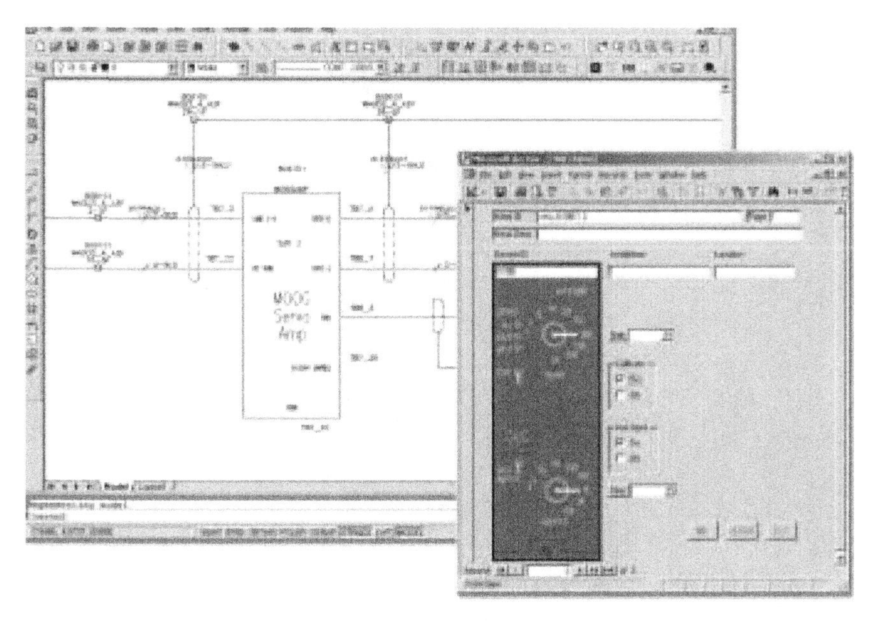

Exdata software links devices in electrical drawing files to additional information stored in an external database. These data can be displayed in a user-friendly, graphical format using database forms.

Easier Analysis With Rocket Science

Analyzing rocket engines is one of Marshall Space Flight Center's specialties. When Marshall engineers lacked a software program flexible enough to meet their needs for analyzing rocket engine fluid flow, they overcame the challenge by inventing the Generalized Fluid System Simulation Program (GFSSP), which was named the co-winner of the NASA Software of the Year award in 2001.

Most rocket analysis tools are either engine or turbopump specific, and some can only analyze one specific engine or perform one part of a test. Consequently, Marshall engineers often use multiple software tools and modules to complete their work. Frequently, however, the different tools cannot communicate, causing roadblocks during analysis. The engineers needed a software program that could plug into virtually any scenario and be operated with minimal training and computer-processing power. As a result, the Marshall development team for GFSSP focused on flexibility when developing the base code for the new software program.

NASA's GFSSP has been extensively verified by comparing its predictions with test data. The software uses a highly intuitive graphical user interface that allows engineers to construct very complex models in a very short time and allows users to model and view the results with a click of a mouse. With changing requirements and needs, the software has evolved to GFSSP version 3.0. This latest version contains a User Subroutine module, making it possible to develop specific applications of the code for various disciplines and customize those applications as needed. As a result, GFSSP has a wide variety of commercial applications in industries that require flow predictions in complex flow circuits.

Concepts NREC, Inc., of White River Junction, Vermont, incorporated GFSSP into its commercially available Cooled Turbine Airfoil Agile Design System (CTAADS). Specializing in all aspects of turbomachinery with applications ranging from aircraft engines to industrial pumps, the company developed the product in conjunction with Harvard Thermal, Inc., of Harvard, Massachusetts, to significantly reduce the total time and cost for designing cooled turbine airfoils.

Gas turbines used for aircraft propulsion must operate at high temperatures for thermal efficiency and power input. However, these extreme temperatures can cause thermal stresses within the turbine blade materials, necessitating that the blades be cooled by air extracted from the engine's compressor for safe and efficient operation. Since thermal efficiency decreases as a result of this extraction, engineers need to understand and optimize the cooling technique, operating conditions, and turbine blade geometry.

An increased understanding of the detailed hot-gas-flow physics within the turbine itself is necessary to design a system that most efficiently cools the turbine

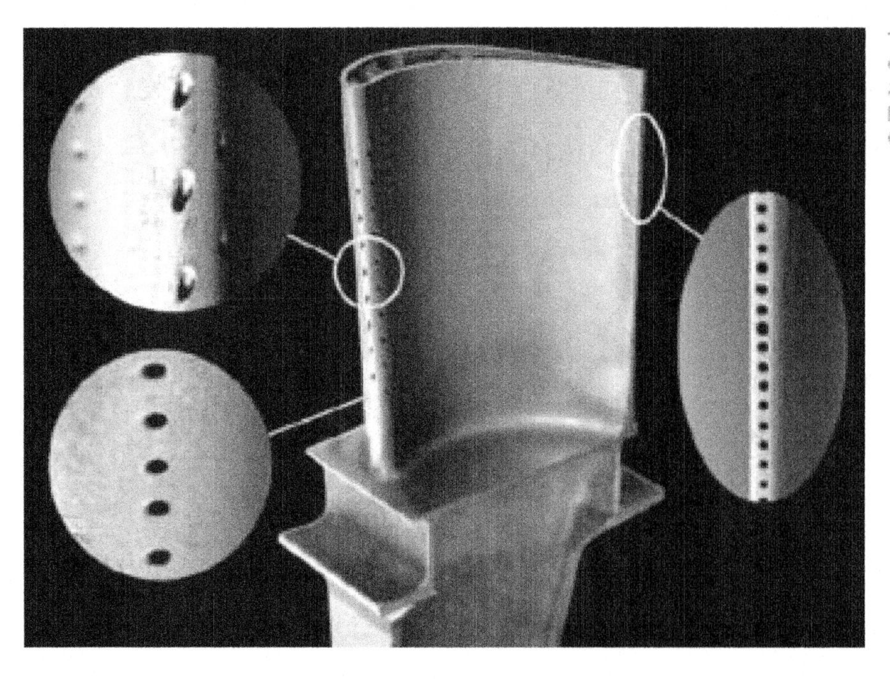

Turbine blades must be properly cooled for safe and efficient operation. This first-stage turbine blade has its cooling holes and drilled trailing edge highlighted.

blades. Since blade life can be reduced by half if the temperature prediction is off by only 50 °F, it is crucial to accurately predict the local heat-transfer coefficient, as well as the local blade temperature, in order to prevent local hot spots and increase turbine blade life.

CTAADS assists users in need of this information by providing a systematic, logical, and rapid three-dimensional (3-D) modeling approach to cooling-system design for cooled axial turbine vanes and blades. The product is a fully integrated suite of independent software modules that supports the rapid generation of airfoil cooling-passage geometry and performs complete 3-D thermal analysis.

CTAADS's internal cooling airflow module, which generates and solves a one-dimensional fluid flow network representing the entire cooling configuration, is built upon GFSSP. Users can construct the fluid flow network with a Concepts NREC-developed graphical user interface. The network is then transformed into a customized input file for GFSSP, and the fluid network solver is a significantly modified and customized version of GFSSP.

The internal cooling airflow module gives users maximum flexibility in defining the fluid flow network. Users can control how many sections are needed to define each cooling passage. The fluid network solver is capable of producing turbine airfoil cooling passage airflow calculations requiring compressible flow with friction, heat addition, and area change; pumping due to blade rotation; and choked flow. Concepts NREC also added numerous resistance options specific to turbine cooling to GFSSP. All options have default correlations for total pressure loss and heat transfer coefficients.

Outside of the success Concepts NREC has found with utilizing GFSSP, other possible commercial applications for the versatile software program include heating ventilation and air conditioning systems, chemical processing, gas processing, power plants, hydraulic control circuits, and various types of fluid distribution systems. Best of all, GFSSP is easy to learn, despite its roots in rocket science. According to Marshall's GFSSP team leader Alok Majumdar, "A goal of ours was to make GFSSP so that an undergraduate engineering student can quickly become proficient with the software."

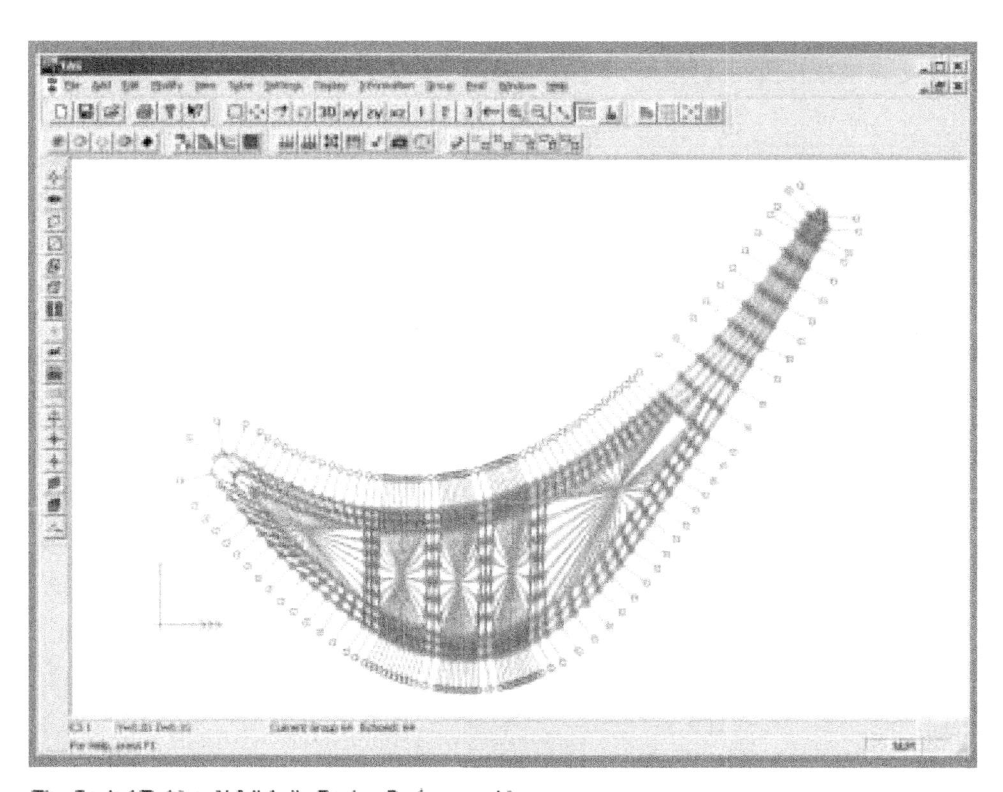

The Cooled Turbine Airfoil Agile Design System provides a systematic, logical, and rapid three-dimensional modeling approach for designing cooled turbine airfoils.

A Tutor That's Up to the Task

Present advances in artificial intelligence (AI) are opening the doors to virtual classroom and training environments where students and trainees are getting undivided attention with one-on-one computer-based instruction. Such new developments are effectively alleviating concerns over long-standing educational issues, such as student-teacher interaction and student-to-teacher ratios. Furthermore, high-level research demonstrates that students who engage in learning using AI-based Intelligent Tutoring Systems (ITS) generally perform better and learn faster, compared to classroom-trained students.

With assistance from NASA's Marshall Space Flight Center, a new breed of ITS for technical training and complex problem-solving has hit the market to provide students and trainees with the decision-making skills necessary to succeed to the next level. The Task Tutor Toolkit™ (T^3), developed by Stottler Henke Associates, Inc., of San Mateo, California, is a generic tutoring system shell and scenario authoring tool that emulates expert instructors and lowers the cost and difficulty of creating scenario-based ITS for technical training.

The functionality of Stottler Henke Associates' T^3 far exceeds that of traditional computer-based training systems, which test factual recall and narrow skills by prompting students to answer multiple-choice or fill-in-the-blank questions. The T^3, on the contrary, lets students assess situations, generate solutions, make decisions, and carry out actions in realistically complex scenarios. At the beginning of each scenario, the T^3 tutoring system presents a briefing that describes the situation and the goals the students should pursue. Each scenario contains a solution template that specifies a partially-ordered sequence of action patterns that match correct sequences of student actions. During each scenario, the built-in simulator notifies the tutoring system of each student action. The T^3 uses this information to evaluate the student action by comparing it with the scenario's solution template and with error rules that detect incorrect actions.

To prevent running into "dead-end" situations, students can request hints and ask questions by clicking on buttons in the user window. Rather than just setting up students with the right answers to complete a scenario, the T^3 simulator explains why the recommended actions should be taken. When students carry out a correct action specified by the solution template, the T^3 awards them credit for applying the principles linked to the action. At the end of each scenario, the tutoring system displays a report card that lists the principles the students successfully—and unsuccessfully—demonstrated. These results

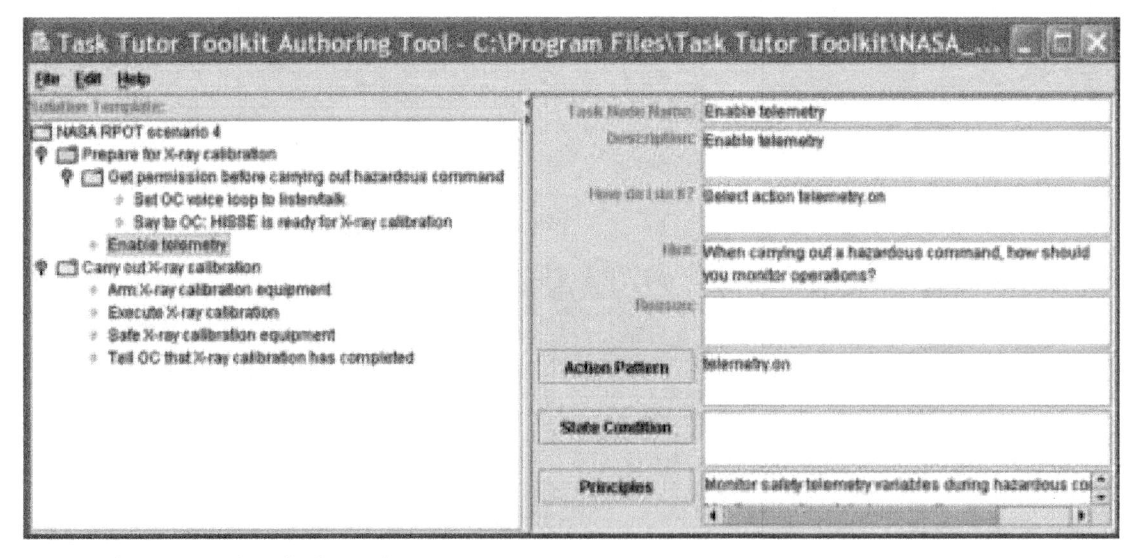

The T^3 Authoring Tool lets instructors edit scenarios by demonstrating, generalizing, and annotating solution templates. Icons in the left pane show actions organized into ordered and unordered groups. The right pane shows the attributes of the selected action, "Enable telemetry."

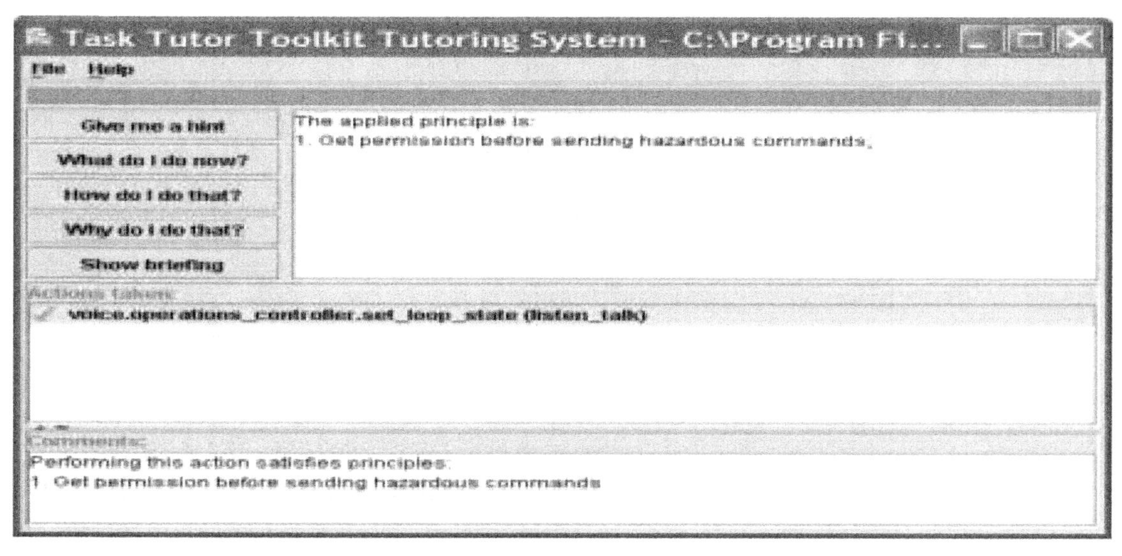

The T³ Tutoring System window provides feedback and hints "on demand" to students during scenarios.

can be sent to a learning management system to record the students' performance, knowledge, and skills.

The T³ is just as easy for instructors and subject matter experts to operate. It consists of a set of Java™ software libraries and applications that let the user create the scenarios quickly and easily, without programming. The system's Simulator and Authoring Tool are used to create application-specific scenarios and lesson plans. The Simulator hosts the correct sequence of actions for the scenario, and the Authoring Tool records these actions to generate the initial solution template.

The tutoring technology was developed from 1997 to 2002, under a **Small Business Innovation Research (SBIR)** contract with Marshall, to address NASA's need for rapid deployment of scenario-based tutoring systems. NASA utilized the toolkit to create the Remote Payload Operations Tutor (RPOT), a system that lets scientists who are new to space mission operations learn to monitor and control their experiments aboard the International Space Station, according to NASA's existing payload regulations, guidelines, and procedures. RPOT combines simulations with automated tutoring capabilities to provide hints and instructional feedback to its students. The system makes it unnecessary for NASA to dedicate an instructor for each student during the simulation exercises.

NASA is exploring the possibility of using the toolkit for onboard training of Space Station flight crews where human instructors cannot be made easily available. This could also serve a purpose in the pre-flight phase when flight crews are traveling away from the high-fidelity training simulators.

Outside of NASA and the corporate and technical training realm, the T³ may be used in educational settings of all levels for math problem-solving and scientific laboratory procedures. The software could also be practical for students at technical institutes and trade schools that teach equipment operations and maintenance.

Software With Strong Ties to Space

With more than 50 years of combined NASA experience under the belts of Tietronix Software, Inc.'s management team, commercial partners from small businesses to Fortune 500 firms are benefiting from the company's landmark achievements in delivering complex, mission-critical systems for NASA's Space Shuttle and International Space Station (ISS) programs.

Recognized by the Houston Business Journal as one of the "Top 25 Software Companies" in the high-tech hotbed of Houston, Texas, Tietronix is a full-service provider of custom software applications and advanced technology solutions. It partners with organizations to help them solve multifaceted business problems via new technology, resulting in increased revenues, productivity, and profitability, and faster time-to-market. Tietronix was founded in 1999 by Victor Tang, Michel Izygon, and Stuart Engelhardt, each having worked on the development of advanced software-based technology solutions, relating to NASA. All three founders worked together at Johnson Space Center, where they developed specialized software to help manipulate the robotic arm of the Space Shuttle.

Tang, the president of Tietronix, is credited with identifying and creating several software tools to support and facilitate astronaut missions, including a flight scheduling automation system, an astronaut activity scheduler, and a robotics display onboard the Shuttle. Izygon, the company's senior vice president and chief technology officer, spent 3 years as a program manager in a technology development contract with Johnson. His duties included management and development of a Web-based electronic workflow system aimed at facilitating and automating the Space Center's mission-critical processes. Engelhardt, Tietronix's vice president, developed software for 5 years at Johnson, authoring various onboard and ground-based mission support applications for Shuttle crews.

Joining the three founders on the management team is Frank Hughes, Tietronix's vice president of training products. Hughes was NASA's Chief of Space Flight Training, where he headed an organization responsible for all Shuttle and Space Station training. He invested more than 33 years at NASA, assisting in the assembly of all U.S. space missions since 1966. Another key management member is French astronaut Jean-Loup Chrétien, recognized as the first Western European to travel in space, as well as the first non-American and non-Soviet to walk in space. Chrétien leads Tietronix's research and development efforts for a new division covering civilian and military applications in the field of optical engineering. As the inventor of an optics-based technology that dramatically improves visibility in extreme sunlight conditions (influenced by his total loss of sight due to sun exposure while in space and while piloting and landing airplanes), Chrétien is spearheading the conception of several prototype products that may one day be used in airplanes, automobiles, and cameras.

Four years after establishment, Tietronix has grown to over 50 experienced engineers and project managers who are dedicated to executing business strategies and applying technologies developed for NASA to commercial markets. The company's TieFlow eProcess software is a paradigm of its successful efforts to transfer technology associated with NASA. Essentially a second-generation workflow system that extends from the founders' work

As a workflow management tool, TieFlow eProcess can automate and simplify any generic or industry-specific work process, helping organizations transform work inefficiencies and internal operations involving people, paper, and procedures into a streamlined, well-organized, electronic-based process.

at Johnson, TieFlow is a simple but powerful business process improvement solution. It can automate and simplify any generic or industry-specific work process, helping organizations to transform work inefficiencies and internal operations involving people, paper, and procedures into a streamlined, well-organized, electronic-based process.

TieFlow increases business productivity by improving process cycle times. The software can expedite generic processes in the areas of product design and development, purchase orders, expense reports, benefits enrollment, budgeting, hiring, and sales. It can also shore up vertical market processes such as claims processing, loan application and processing, health care administration, contract management, and advertising agency traffic. The processes can be easily and rapidly captured in a graphical manner and enforced together with rules pertaining to assignments that need to be performed. Aside from boosting productivity, TieFlow also reduces organizational costs and errors.

TieFlow was developed with **Small Business Innovation Research (SBIR)** assistance from Johnson. The SBIR support entitles all Federal Government agencies to utilize the TieFlow software technology free of charge. Tietronix emphasizes that TieFlow is an outstanding workflow resource that could produce dramatic productivity and cost improvements for all agencies, just as it has done and continues to do for NASA. The Space Agency is currently using the software throughout several mission-critical offices, including the Mission Operations Directorate and the Flight Director's Office, for worldwide participation of authorized users in NASA processes. At the Flight Director's Office, TieFlow allows personnel to electronically submit and review changes to the flight rules carried out during missions.

Outside of government, Tietronix secured a commercial contract to implement the TieFlow technology into a vertical solution for the health care industry. The Home Care Connect™ Web-based Point of Care Solution workflow tool was developed in cooperation with Inter-Active Healthcare, Inc., a leading home care agency in the southwest. With its simple Web interface, field professionals can collect and validate Outcome and Assessment Information Set (OASIS) and clinical data from the start of care through the time the patient is discharged, all without having to install any software. The tool is also capable of working with any back-office claims processing and billing system, as well as capturing and transferring data from telemedicine equipment.

Tietronix also offers a Virtual Tour software product that provides users with a "total immersion experience." Unlike traditional interactive multimedia products, the Virtual Tour technology creates a complete suspension of disbelief that lets users travel freely through a virtual space. For architects and engineers, the software can create extremely detailed virtual structures that they can "walk" through to head off any potential problems prior to starting a construction project. It is also ideal for emergency and operational facility training purposes, travel and tourism destinations, commercial and residential real estate sales, and science teaching. The software has been utilized to create a virtual interactive tour of Johnson Space Center, a "fly-through" tour of the ISS, and a walk-through tour of the multi-story Hilton Clear Lake Hotel in Houston, complete with automatic doors and operational elevators.

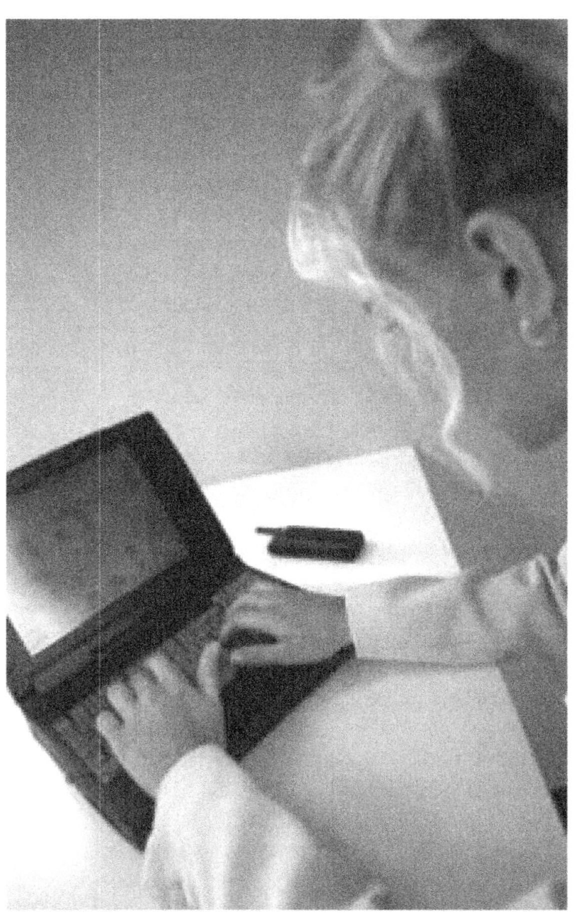

Using a laptop and the Internet, Home Care Connect™ enables field professionals to validate Outcome and Assessment Information Set and clinical data from the start of care through the time the patient is discharged, all without having to install any software.

Home Care Connect™ is a trademark of Tietronix Software, Inc.

Image Acquisition in Real Time

In 1995, Carlos Jorquera left NASA's Jet Propulsion Laboratory (JPL) to focus on erasing the growing void between high-performance cameras and the requisite software to capture and process the resulting digital images. Since his departure from NASA, Jorquera's efforts have not only satisfied the private industry's cravings for faster, more flexible, and more favorable software applications, but have blossomed into a successful entrepreneurship that is making its mark with improvements in fields such as medicine, weather forecasting, and X-ray inspection.

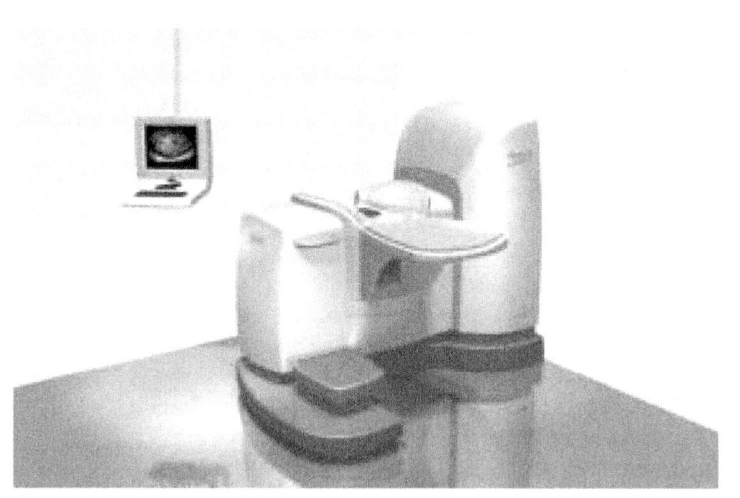

Image courtesy of Advanced Imaging Technologies, Inc.

AcquireNow is the software imaging engine within the Avera™ Breast Imaging System.

Formerly a JPL engineer who constructed imaging systems for spacecraft and ground-based astronomy projects, Jorquera is the founder and president of the three-person firm, Boulder Imaging Inc., based in Louisville, Colorado. Joining Jorquera to round out the Boulder Imaging staff are Chief Operations Engineer Susan Downey, who also gained experience at JPL working on space-bound projects—including Galileo and the Hubble Space Telescope, and Vice President of Engineering and Machine Vision Specialist Jie Zhu Kulbida, who has extensive industrial and research and development experience within the private sector.

The Boulder Imaging team's vast engineering talent shines through on its flagship imaging capture and processing software product, *AcquireNow*. As a Component Object Model (COM) component, *AcquireNow* provides software developers with the programmatic tools necessary to create software applications that can acquire images from high data rate digital cameras and apply image-processing algorithms to those images (COM is a software architecture developed by Microsoft® that allows components made by different software vendors to be combined into a variety of applications). Essentially, *AcquireNow* users can obtain images from any camera, using any frame grabber board, without having to write any hardware-specific software.

The most important feature of *AcquireNow* is its ability to perform sustained image acquisition from high frame rate and/or high-resolution digital cameras. With the correct pairing of camera and frame grabber hardware, and a dual peripheral component interconnect bus system (an expansion slot for personal computers), the software can acquire image data at rates exceeding 500 megabytes per second. The technology also encompasses powerful image-processing algorithms, including averaging, flat fielding, scaling, edge detection, blob analysis, and display. Additionally, it uses multiple threads of execution to allow image processing to occur in one thread, while an acquisition thread is waiting for a frame to come in from the camera. This multi-threaded option permits the software to take advantage of multiple computers, thus, increase performance on these systems.

The *AcquireNow* software package includes the *AcquireNowClient* stand-alone application, which can be used to obtain, display, and save images to disk. The source code for *AcquireNowClient* is also included, so that customers may freely use it as a base for internal or commercial applications.

In 2002, Advanced Imaging Technologies, Inc., of Preston, Washington, licensed *AcquireNow* and embedded it in its Avera™ Breast Imaging System, which permits rapid assessment of breast tissue in real time, without the discomfort of compression or the risk of harmful radiation. The Avera Breast Imaging System features Advanced Imaging Technologies' first commercial use of the company's patented Diffractive Ultrasound innovation, a unique imaging technology based on the principles of sound. Unlike conventional ultrasound that relies on sound's reflective properties, Diffractive Ultrasound takes advantage of the diffractive properties to collect high-resolution images from soft tissue structures, such as the breast. *AcquireNow* is responsible for capturing, processing, and depicting the images clearly on a computer screen.

AcquireNow has also been instrumental in the detection of ice crystals that form in high-altitude clouds. Boulder Imaging developed both the real-time processing system and the post-processing software for Boulder, Colorado-based Stratton Park Engineering Company (SPEC), Inc.'s Cloud Particle Imager, a device that takes high-resolution digital images of cloud

Boulder Imaging Inc., developed both the real-time processing system and the post-processing software for Stratton Park Engineering Company, Inc.'s Cloud Particle Imager (mounted on the underbelly of the Learjet).

Image courtesy of Stratton Park Engineering Company, Inc.

particles and flew aboard NASA's WB-57 aircraft during the Cirrus Regional Study of Tropical Anvils and Cirrus Layers – Florida Area Cirrus Experiment (CRYSTAL-FACE) mission.

SPEC's Cloud Particle Imager is comprised of a high-speed camera that snaps extremely fast pictures of the particles inside of a cloud as the plane flies through it. *AcquireNow* is used to detect the edges of ice particles in real time, and determine what in an image is a particle and what is not. Sizing information for each ice particle identified in the images is computed, and statistical information is generated, tracking particle size distribution over time. Researchers can then use this data to study the microscopic factors that influence cloud physics, and ultimately, how clouds affect the atmosphere's temperature. The Cloud Particle Imager—also popular among international governments interested in cloud research related to global climate change—is just one of seven SPEC products to incorporate the *AcquireNow* technology.

AcquireNow is also making its presence known on manufacturing assembly lines, aiding in the quick detection of faulty circuit boards and other machine parts. Agilent Technologies, of Palo Alto, California, purchased *AcquireNow* for use in its 5DX quality control system, designed to detect structural defects through an X-ray source.

More recently, Boulder Imaging embarked on a project in which *AcquireNow* is being used to image the conditions of highways. By linking the software with a global positioning system and a special camera that takes multiple shots of a road surface at 50 miles per hour, *AcquireNow* can portray images of the road structure that prove valuable to highway engineers.

The company notes that future applications for the software may be seen in airports to expedite and enhance baggage searches, in machine vision systems to read zip codes on mailing labels, and in the entertainment industry for digital data storage.

Microsoft® is a registered trademark of Microsoft Corporation.
Avera™ is a trademark of Advanced Imaging Technologies, Inc.

Image courtesy of Samsung SDS America.

AcquireNow is featured in uniAMS, a commercial product of Samsung SDS America, used for measuring pavement conditions.

Free-Flowing Solutions for CFD

Licensed to over 1,500 customers worldwide, an advanced computational fluid dynamics (CFD) post-processor with a quick learning curve is consistently providing engineering solutions, with just the right balance of visual insight and hard data.

FIELDVIEW™ is the premier product of JMSI, Inc., d.b.a. Intelligent Light, a woman-owned, small business founded in 1994 and located in Lyndhurst, New Jersey. In the early 1990s, Intelligent Light entered into a joint development contract with a research-based company to commercialize the post-processing FIELDVIEW code. As Intelligent Light established itself, it purchased the exclusive rights to the code, and structured its business solely around the software technology. As a result, it is enjoying profits and growing at a rate of 25 to 30 percent per year.

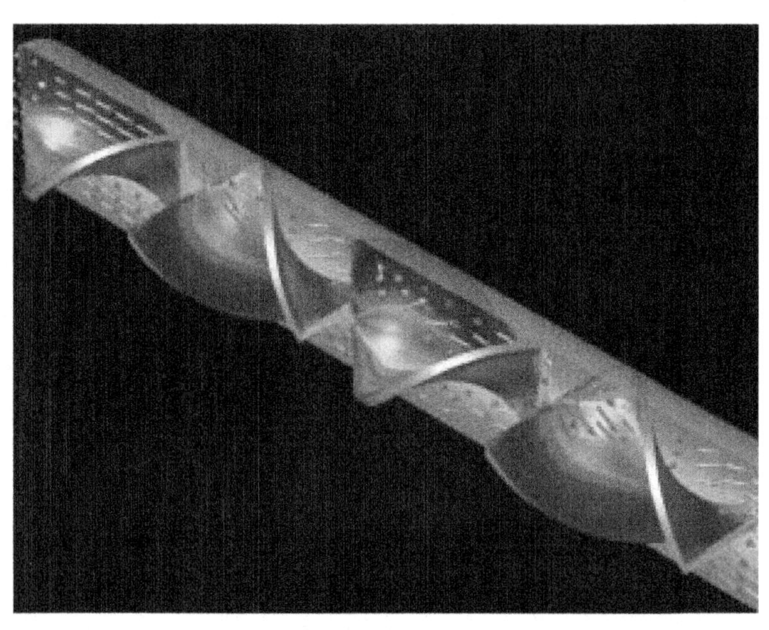

FIELDVIEW™ 8's advanced programmability function successfully calculated the shear stress behavior on the blades of a Kenics® static mixer, a tool that is widely used in the food and chemical industries for in-line blending of liquids.

Advancements made from the earliest commercial launch of FIELDVIEW, all the way up to the recently released versions 8 and 8.2 of the program, have been backed by research collaboration with NASA's Langley Research Center, where some of the world's most progressive work in transient (also known as time-varying) CFD takes place. In 1994, Intelligent Light was contracted by Langley to develop tools for a special case of aero-acoustic post-processing. The company successfully delivered a modified version of FIELDVIEW to a NASA contracting branch that performs numerical integration of scalar functions over surfaces that are "swept" through time.

The following year, Langley awarded Intelligent Light a **Small Business Innovation Research (SBIR)** contract for advanced research into visualization technology. The SBIR focused on developing sophisticated transient data-handling capabilities to be integrated into the FIELDVIEW product. According to one Langley researcher who provided the data set during Phase I of the contract, the prototype demonstrated "an enabling technology that gives NASA a unique, affordable solution to a very difficult post-processing problem." The Phase II follow-on effort expanded the scope of the new data access and visualization technologies to unstructured and hybrid grids, as well as analyses that combine fluid/structure interaction. The research and development resulting from both phases of the SBIR project was implemented into the software in 1996, alleviating the complications of handling very large transient or steady-state data sets for all users.

To keep in step with technological progression in the CFD field, Intelligent Light once again teamed up with Langley in 1999 to create a next-generation platform that would permit design options to be evaluated with greater speed and efficiency. Going into the SBIR contract, Intelligent Light contended that CFD post-processing software was either too visual, lacking the hard data for decision-making, or too quantitative, with an overabundance of numbers that were poorly illustrated. This time, the SBIR focused on advanced data query techniques to the extremely large data sets that were becoming common in the CFD environment and, therefore, overburdening users by causing backlogs.

Intelligent Light's SBIR work aimed to correct these problems and was eventually commercialized in FIELDVIEW 8. By working in conjunction with NASA and other top CFD authorities, the company was able to deliver a new form of integrated post-processing automation, enabling users to quickly understand flow patterns, automate quantitative analyses, and deliver convincing, accurate presentations.

With FIELDVIEW 8, the company boasts a new advanced programmability function called FVX™ that lets users read data sets, create and manipulate surfaces, and perform complex quantitative post-processing tasks using a real programming language. For example, FVX was used to successfully calculate the shear stress behavior on the blades of a Kenics® static mixer, a tool that is widely used in the food and chemical industries for in-line blending of liquids. This information is often necessary when determining whether or not the mixing process will exceed the allowable maximum shear stress for the materials being combined. The results generated by FVX were plotted into a histogram indicating flow direction and points where higher stress would degrade the materials.

The latest adaptation of FIELDVIEW, version 8.2 (a derivative of version 8 intended to address smaller scale processing), features improved data set comparison, which permits easy visual and numerical comparison between two or more data sets whose grids are topologically similar. A "Dataset Comparison" mode in the software's built-in CFD calculator allows for the creation of formulas that may reference quantities of more than one data set. For example, an engineer can quickly plot the difference in temperature at the outlet of different designs. Another key upgrade is the addition of blended transparency, enabling users to create the highest quality representations and animations. Additionally, the enhanced 8.2 version offers many new capabilities to help designers and engineers produce better results in less time.

FIELDVIEW is widely used in the aerospace, automotive, defense, and manufacturing sectors. Lockheed Martin, for example, has standardized on the technology for CFD post-processing needs. The company employs FIELDVIEW to design tactical aircraft systems, including F-16, F-22, and F-35 Joint Strike Fighters. Boeing also makes use of the software in both its commercial and defense sectors.

Firms such as Pratt & Whitney, General Electric, Rolls Royce, and Honeywell have adopted FIELDVIEW to design state-of-the-art gas turbine engines. Moreover, auto makers such as the Ford Motor Company, Honda, and Toyota utilize it in applications such as under-hood cooling and airflow; cabin heating, venting, and air conditioning; powertrain design; glass fabrication; and paint room design. Ford, among others, has entered into a worldwide corporate licensing agreement with Intelligent Light which provides for software licenses, training, and customization services.

In addition to Langley Research Center, FIELDVIEW is widely used at Ames and Glenn Research Centers, Johnson Space Center, and Marshall Space Flight Center. Intelligent Light continues to work with its largest customers, including NASA, to find new ways to contend with the mountains of data resulting from CFD computations.

FIELDVIEW™ and FVX™ are trademarks of IMSI, Inc., d.b.a. Intelligent Light.
Kenics® is a registered trademark of Chemineer, Inc.

FIELDVIEW™ is used to design tactical aircraft systems, including F-16, F-22, and F-35 Joint Strike Fighters.

Efficient, Multi-Scale Designs Take Flight

Engineers can solve aerospace design problems faster and more efficiently with a versatile software product that performs automated structural analysis and sizing optimization. Collier Research Corporation's HyperSizer® Structural Sizing Software is a design, analysis, and documentation tool that increases productivity and standardization for a design team. Based on established aerospace structural methods for strength, stability, and stiffness, HyperSizer can be used all the way from the conceptual design to in-service support.

The software originated from NASA's efforts to automate its capability to perform aircraft strength analyses, structural sizing, and weight prediction and reduction. With a strategy to combine finite element analysis with an automated design procedure, NASA's Langley Research Center led the development of a software code known as ST-SIZE from 1988 to 1995. Collier Research employees were principal developers of the code along with Langley researchers. The code evolved into one that could analyze the strength and stability of stiffened panels constructed of any material, including light-weight, fiber-reinforced composites.

After obtaining an exclusive NASA license for ST-SIZE in 1996, Hampton, Virginia-based Collier Research combined the code with other company proprietary software, taking it from the research level to the commercial level with full documentation, training, and quick response support. Marketed as HyperSizer, the software couples with commercial finite element analyses to enable system level performance assessments and weight predictions; conceptual and preliminary design optimization of material selection/layup and structural member sizing; and structural failure analysis and automated stress reports.

Continuing to advance HyperSizer by expanding its capabilities and features, Collier Research began working with NASA's Glenn Research Center in 2001. The goal, initiated through several NASA contracts and grants, was to integrate and commercialize the Micromechanics Analysis Code with Generalized Method Cells (MAC/GMC) and the higher-order theory for functionally graded materials (HOTFGM). Both MAC/GMC and HOTFGM were developed collaboratively with personnel at Glenn, the University of Virginia, the Ohio Aerospace Institute, and Israel's Tel-Aviv University.

MAC/GMC is a well-documented software package for the design and analysis of advanced composite materials that accurately predicts the elastic and inelastic thermomechanical response of multiphased materials including polymer-, ceramic-, and metal-matrix composites. The software enables engineers and material scientists to design and analyze composite materials for a given application accurately and easily. The well-established generalized method of cells micromechanics theory provides the underlying analysis technology for MAC/GMC.

The HOTFGM theory and code are designed to model the thermal and mechanical response of structures with arbitrary cross-sections composed of functionally graded materials. These new technologies enable the HyperSizer analysis to localize beyond its traditional stiffened panel and laminate ply scale. Equipped with the MAC/GMC and HOTFGM capabilities, HyperSizer can now design on the microscale, considering the individual fiber and matrix phases, their arrangements, and their likelihood to initiate failure of the global aerospace structure.

Through its efforts to commercialize HOTFGM and MAC/GMC, Collier Research now includes Hyper-FGM

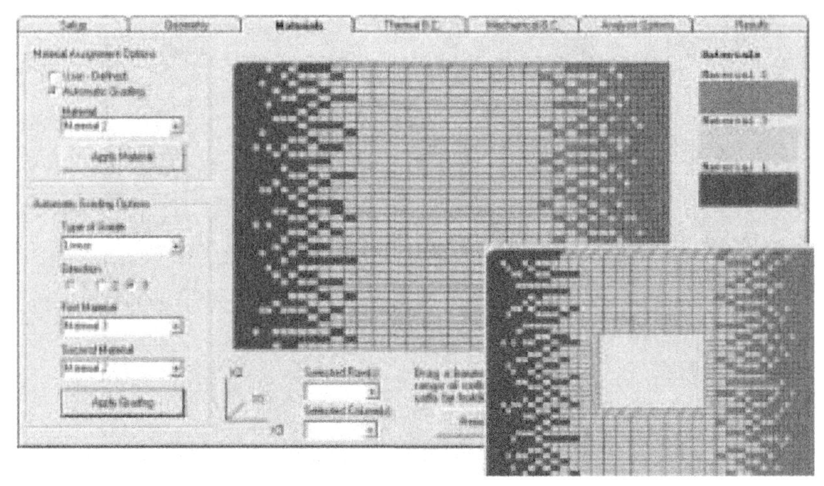

Collier Research Corporation's Hyper-FGM software package is a tool for the design and analysis of functionally graded materials. It includes pre- and post-processing through an intuitive graphical user interface, along with the HOTFGM thermo-mechanical analysis engine.

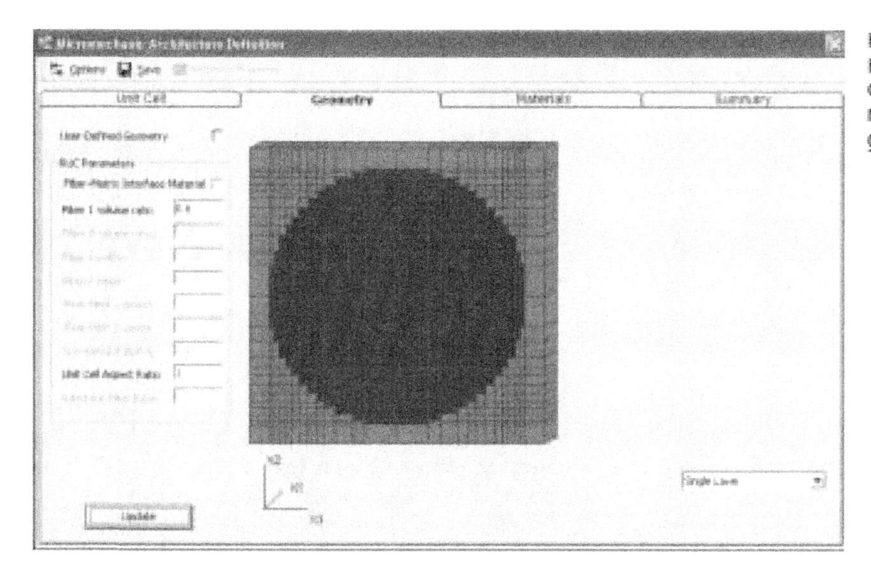

Hyper-MAC, an add-on module for the HyperSizer® software, enables users to create and use new homogeneous materials based on micromechanics-generated effective properties.

and Hyper-MAC as part of its product line. Hyper-FGM uses HOTFGM as its underlying analysis technology and is a self-contained software package developed as a tool for the design and analysis of functionally graded materials (FGMs). It includes pre- and post-processing through an intuitive graphical user interface, along with the HOTFGM thermo-mechanical analysis engine.

The Hyper-FGM interface builds on the well-established accuracy of the HOTFGM mechanics methodology, making HOTFGM available as a practical commercial product. After an FGM problem is specified, the HOTFGM-based analysis can be executed from within Hyper-FGM, and the results are automatically loaded into the interface for viewing and post-processing. With its ease of use, Hyper-FGM provides an effective alternative to the more user-intensive approach of modeling FGMs through finite element analysis packages.

Hyper-MAC is an add-on module for the HyperSizer software that combines the constituent buildup (formation of repeating unit cell) and nonlinear analysis of the NASA-developed MAC/GMC with HyperSizer's data integrity, established laminate analysis methodology, and material property editing interface. The inclusion of micromechanics within HyperSizer creates a new distinction between material types: heterogeneous and homogeneous. Homogeneous materials have no underlying micromechanics architecture and are analyzed using only equivalent, or effective, properties. Heterogeneous materials have effective properties that are calculated using a micromechanics model given specified microscale constituent properties and architecture. Consequently, Hyper-MAC enables users to create and use new homogeneous materials based on micromechanics-generated effective properties and to perform virtual experiment simulations to determine the effective response of heterogeneous materials.

While targeted to the aerospace industry, the line of HyperSizer products impacts a variety of additional commercial applications. In the transportation industry, HyperSizer's cost-effective design capability with stiffened wall construction concepts makes it ideal for designing and analyzing rail cars, tractor trailers, dump trucks, and shipping containers. HyperSizer can also be used for new and innovative designs for the marine industry, ranging from scantlings of composite hull panels for high speed light craft to stiffened plate panels of hull girders, side shells, and bottom structures. Since HyperSizer can achieve the lightest weight for any marine design, it benefits high-speed vessels such as passenger catamarans and rescue vehicles that are weight optimized to obtain high speeds.

Monte Carlo Methodology Serves Up a Software Success

Widely used for the modeling of gas flows through the computation of the motion and collisions of representative molecules, the Direct Simulation Monte Carlo method has become the "gold standard" for producing research and engineering predictions in the field of rarified gas dynamics. Direct Simulation Monte Carlo was first introduced in the early 1960s by Dr. Graeme Bird, a professor at the University of Sydney, Australia. It has since proved to be a valuable tool to the aerospace and defense industries in providing design and operational support data, as well as flight data analysis.

In 2002, NASA brought to the forefront a software product that maintains the same basic physics formulation of Dr. Bird's method, but provides effective modeling of complex, three-dimensional, "real" vehicle simulations and parallel processing capabilities to handle additional computational requirements, especially in areas where computational fluid dynamics (CFD) is not applicable. NASA's Direct Simulation Monte Carlo Analysis Code (DAC) software package is now considered the Agency's premier high-fidelity simulation tool for predicting vehicle aerodynamics and aerothermodynamic environments in rarified, or low-density, gas flows.

Additionally, DAC was recognized as the co-winner of the prestigious 2002 NASA Software of the Year award, for its time- and money-saving abilities. Because tests to acquire information on the interactions of spacecraft and rarified environments are difficult and expensive to perform, NASA believes DAC has the potential to save millions of dollars. NASA also acknowledges that DAC is easier to use and 100 times faster than the "standard process" software it replaced.

Developed by an engineering team in Johnson Space Flight Center's Aeroscience and Flight Mechanics Division, DAC models the flow of low-density gases over flight surfaces. The accurate modeling of these flows, for which experimental facilities are virtually nonexistent, is critical for the protection of valuable NASA assets, and for ensuring crew safety and overall mission success. The software supports numerous Johnson programs, including the International Space Station (ISS), the X-38, and Space Shuttle servicing missions to the Hubble Space Telescope. DAC is also in use at other field centers, to monitor the Mars Pathfinder, Stardust, Genesis, X-33, X-37, Mars Global Surveyor, and Mars Odyssey vehicles.

More broadly, the DAC software serves as the benchmark for predicting flowfields generated on orbit by the Space Shuttle reaction control system thrusters, and the resultant impingement (effects of thruster firing when one spacecraft approaches another) on the ISS. Use of the DAC code in predicting flowfields has led to significant changes in docking procedures and venting operations. The technology is further used to provide information to help optimize and verify maneuvers of the Mars-orbiting spacecraft after they were slowed by repeatedly skimming through the planet's atmosphere ("aerobraking"), instead of depending upon thrusters for deceleration. This technique enabled the spacecraft to be lighter, which in return reduced launch costs.

The generality incorporated into DAC has allowed for widespread use of the software within the aerospace community beyond NASA. The National Missile Defense embraced the technology for interception of hostile missiles launched into the upper atmosphere, beyond the limits where traditional CFD methods would

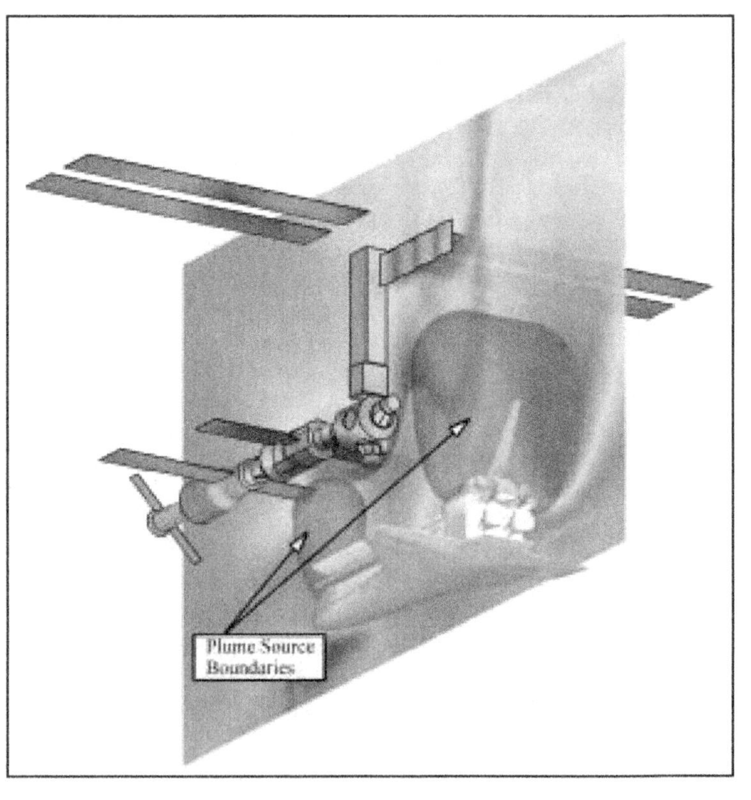

NASA's Direct Simulation Monte Carlo Analysis Code serves as the benchmark for predicting flowfields generated on orbit by the Space Shuttle reaction control system thrusters, and the resultant impingement on the International Space Station.

be functional. Department of Defense contractors such as Boeing and Raytheon employ the software to assess candidate kinetic warheads for ballistic missile defense systems. These contributions to missile defense are believed to bolster homeland security and provide citizens with an improved sense of safety.

DAC research is also prevalent at universities all across the country. The University of Colorado was first in utilizing DAC to analyze the rarified aerodynamics of a "wave rider" (a hypersonic vehicle that glides on its own shockwave) using the method of osculating cones. This method was developed at the University of Colorado in the late 1980s, and is now the standard for wave rider design. The motivation behind this research is the potential for using wave riders for aero-maneuvers in planetary atmospheres, particularly for aero-gravity assists. Such maneuvers were first proposed by Jet Propulsion Laboratory engineers approximately 15 years ago.

At the University of Maryland, DAC has been used to explore various problems of direct relevance to NASA and the U.S. Air Force. One instance involves the behavior of a satellite in a so-called "dipping orbit," which includes a low-altitude pass through the Earth's upper atmosphere. Several proposed future space missions will fly dipping orbits to perform experiments in the upper atmosphere, including Goddard Space Flight Center's Geospace Electrodynamic Connections mission.

Another project used DAC as a baseline to study the aerodynamic environment at the leading edge of a spacecraft as it reenters from orbit. University faculty and students are currently exploring options for minimizing, or even eliminating, the communications blackout period of a reentering vehicle. Field experts believe that, through careful selection of the spacecraft geometry, the amount of high-temperature plasma surrounding a reentering spacecraft can be dramatically reduced, permitting radio waves to travel more easily from the spacecraft to ground control, and back. Overall, University of Maryland researchers have found DAC to be extremely powerful, computationally efficient, and very easy to install and use.

The unique flow-solvers adapted to DAC make the software a good fit for other applications in which the object within the flowfield is extremely small, such as micro-electromechanical and nanotechnology devices. The software technology is also expected to make an impact in materials processing, including chemical vapor deposition and etching of thin films, and internal transport analysis of outgassing contaminants.

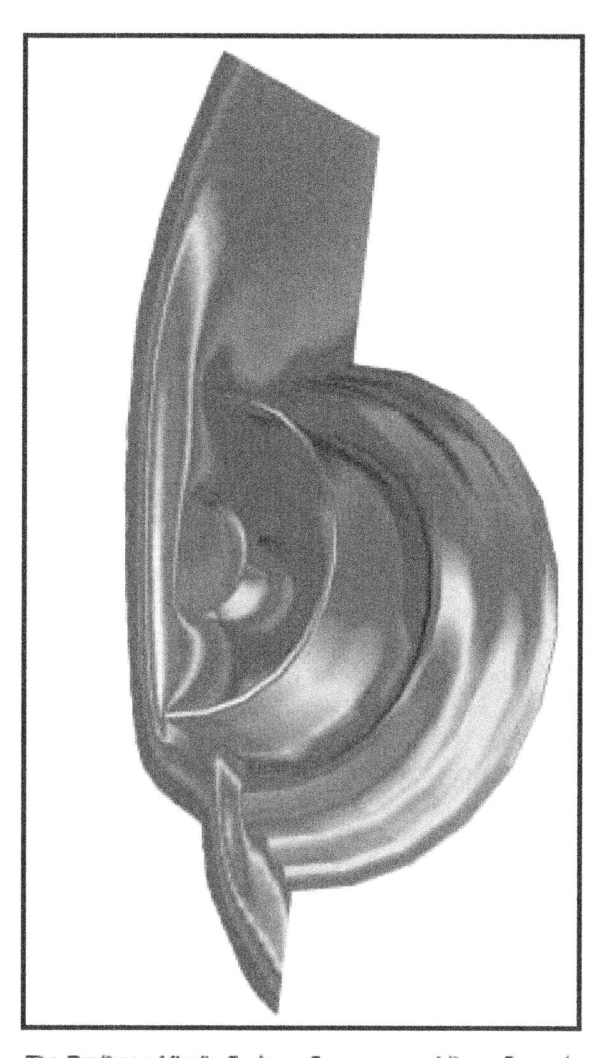

The Raytheon Missile Systems Company used the software to develop Standard Missile-3 Infrared Seekers for the Missile Defense Agency. The seekers successfully intercepted a test target vehicle in January 2002, marking the first ship-launched ballistic missile intercept.

Building Safer Systems With SpecTRM

System safety, an integral component in software development, often poses a challenge to engineers designing computer-based systems. While the relaxed constraints on software design allow for increased power and flexibility, this flexibility introduces more possibilities for error. As a result, system engineers must identify the design constraints necessary to maintain safety and ensure that the system and software design enforces them.

Safeware Engineering Corporation, of Seattle, Washington, provides the information, tools, and techniques to accomplish this task with its Specification Tools and Requirements Methodology (SpecTRM). NASA assisted in developing this engineering toolset by awarding the company several **Small Business Innovation Research (SBIR)** contracts with Ames Research Center and Langley Research Center. The technology benefits NASA through its applications for Space Station rendezvous and docking.

SpecTRM aids system and software engineers in developing specifications for large, complex safety-critical systems. The product enables engineers to find errors early in development so that they can be fixed with the lowest cost and impact on the system design. SpecTRM traces both the requirements and design rationale (including safety constraints) throughout the system design and documentation, allowing engineers to build required system properties into the design from the beginning, rather than emphasizing assessment at the end of the development process when changes are limited and costly.

Engineers ensure that software specifications possess desired safety properties through manual inspection, formal analysis, simulation, and testing. SpecTRM provides support for all of these activities. The simulation of specifications in SpecTRM graphically illustrates the behavior of software from the requirements model, allowing the software requirements to be tested and validated before the costly process of generating design and code. SpecTRM's specification slicing tool cuts through even the most complex systems to assist reviewers in validating requirements by making the most important system behavior stand out.

As a bridge among diverse groups of system, software, and safety engineers, SpecTRM facilitates communication and the coordinated design of components and interfaces. The product's executable requirements specification language can be easily read and reviewed by all system engineering disciplines, helping to provide seamless transitions and mappings between the various development and maintenance stages.

SpecTRM is based on proven research methods in flight management systems, air traffic control systems, and the Traffic Alert and Collision Avoidance System. The tool ensures these methods and analyses are robust, user-friendly, and automated to the point that they can be used in an industrial setting, benefiting the aerospace and transportation industries. SpecTRM can also be applied to designs for automotive systems, defense systems, and medical devices. Safeware Engineering Corporation offers comprehensive consulting and training services to new and existing SpecTRM users, helping customers to meet their system specification and design needs.

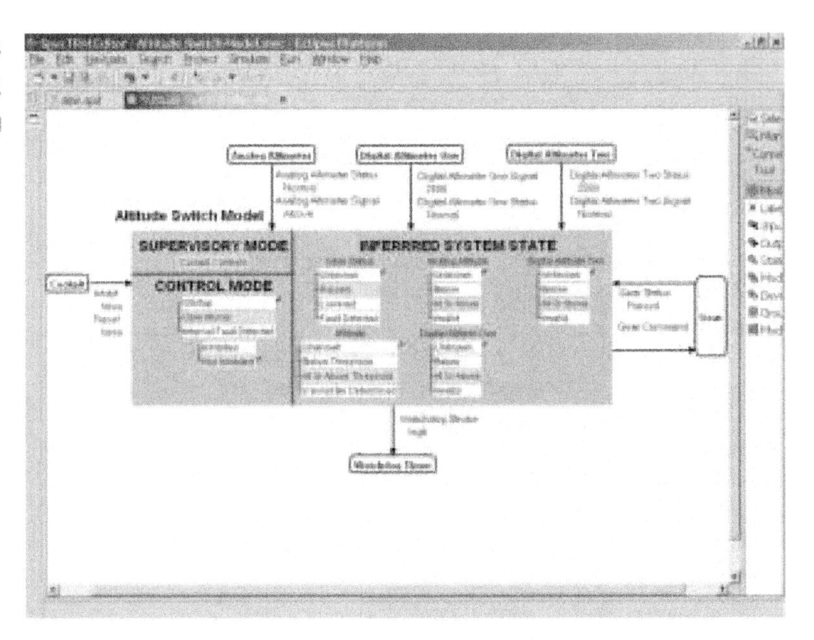

The Specification Tools and Requirements Methodology provides the information, tools, and techniques engineers need to identify design constraints for system safety during software development.

Reconfigurable Hardware Adapts to Changing Mission Demands

A new class of computing architectures and processing systems, which use reconfigurable hardware, is creating a revolutionary approach to implementing future spacecraft systems. With the increasing complexity of electronic components, engineers must design next-generation spacecraft systems with new technologies in both hardware and software. Derivation Systems, Inc., of Carlsbad, California, has been working through NASA's **Small Business Innovation Research (SBIR)** program to develop key technologies in reconfigurable computing and Intellectual Property (IP) soft cores.

Founded in 1993, Derivation Systems has received several SBIR contracts from NASA's Langley Research Center and the U.S. Department of Defense Air Force Research Laboratories in support of its mission to develop hardware and software for high-assurance systems. Through these contracts, Derivation Systems began developing leading-edge technology in formal verification, embedded Java,™ and reconfigurable computing for its PF3100,™ Derivational Reasoning System (DRS™), FormalCORE IP,™ FormalCORE PCI/32,™ FormalCORE DES,™ and LavaCORE™ Configurable Java Processor, which are designed for greater flexibility and security on all space missions.

The PF3100 is an ultra high-density reconfigurable module using Xilinx® Virtex®-II Platform Field Programmable Gate Arrays (FPGAs) in the rugged, compact, industry standard PC/104+ form factor. The module is a multi-function, single-inventory device. Hardware algorithms can be stored in on-board Flash memory or downloaded from a host system to configure the module for a specific mission requirement. These algorithms can be dynamically loaded onto the FPGAs to change the PF3100's function.

Derivation Systems developed the DRS through a 1994 Langley SBIR contract. The system allows an engineer to develop hardware algorithms using formal verification methods. Upon receiving a subsequent SBIR contract with Langley, the company adapted DRS as the underlying technology for its FormalCORE IP product line to develop a library of IP cores targeted to reconfigurable hardware.

FormalCORE IP components are pre-defined, pre-verified system functions in the form of a netlist that serves as building blocks for new designs. Design teams can quickly incorporate these building blocks into their designs while maintaining a high degree of assurance from formal verification tools. Engineers within the computer, networking, and semiconductor markets can manipulate these pre-designed components to develop reusable designs with reduced errors and design cycle times, as well as reduced time-to-market for electronic products and systems.

The FormalCORE IP family includes FormalCORE PCI/32—a 32-bit/33Mhz PCI interface core, FormalCORE DES—an implementation of the DES encryption algorithm, and the LavaCORE Configurable Java Processor—a 32-bit processor that executes Java byte code directly in hardware.

With the introduction of the PF3100 and its FormalCORE IP technology, Derivation Systems has launched itself as a leading supplier of PC/104-based FPGA boards and IP for the embedded systems market. The company ships its products to a variety of aerospace, government, telecommunications, and industrial customers. The PF3100 has been adopted to provide a comprehensive solution to a broad range of applications including data acquisition, hyper-spectral imaging, engine control, three-dimensional bio-imaging, software radio, robotics, power conditioning, telecommunications, and prototyping fault-tolerant bus architectures.

Derivation Systems' customers include the U.S. Air Force, Northrop Grumman, Ericsson, General Dynamics, Lawrence Livermore National Laboratory, Lockheed Martin Corporation, and NASA.

Java™ is a trademark of Sun Microsystems, Inc.
PF3100,™ DRS,™ FormalCORE IP,™ FormalCORE PCI/32,™ FormalCORE DES,™ and LavaCORE™ are trademarks of Derivation Systems, Inc.
Xilinx® and Virtex® are registered trademarks of Xilinx, Inc.

Development of Derivation Systems, Inc.'s PF3100,™ an ultra high-density reconfigurable module, was supported in part by funding from Langley Research Center and Air Force Research Laboratories.

The Logical Extension

The same software controlling autonomous and crew-assisted operations for the International Space Station (ISS) is enabling commercial enterprises to integrate and automate manual operations, also known as decision logic, in real time across complex and disparate networked applications, databases, servers, and other devices, all with quantifiable business benefits.

Auspice® Corporation, of Framingham, Massachusetts, developed the Auspice TLX® (The Logical Extension) software platform to effectively mimic the human decision-making process. Auspice TLX automates operations across extended enterprise systems, where any given infrastructure can include thousands of computers, servers, switches, and modems that are connected, and therefore, dependent upon each other.

The concept behind the Auspice software spawned from a computer program originally developed in 1981 by Cambridge, Massachusetts-based Draper Laboratory for simulating tasks performed by astronauts aboard the Space Shuttle. At the time, the Space Shuttle Program was dependent upon paper-based procedures for its manned space missions, which typically averaged 2 weeks in duration. As the Shuttle Program progressed, NASA began increasing the length of manned missions in preparation for a more permanent space habitat. Acknowledging the need to relinquish paper-based procedures in favor of an electronic processing format to properly monitor and manage the complexities of these longer missions, NASA realized that Draper's task-simulation software could be applied to its vision of year-round space occupancy.

In 1992, Draper was awarded a NASA contract to build User Interface Language software to enable autonomous operations of a multitude of functions on Space Station Freedom (the station was redesigned in 1993 and converted into the international venture known today as the ISS). The resulting software was certified by NASA and installed on the Space Station's Command and Control and Payload Control Processors, through Marshall Space Flight Center and Johnson Space Center.

Known as Timeliner, the software was used for the first time aboard the ISS in June 2003 to autonomously activate and control experiment payloads in the Microgravity Science Glovebox, a sealed container with built-in gloves that provides a safe, enclosed workspace for investigations conducted in the low-gravity environment. It is further anticipated to automate other ISS procedural tasks typically performed by human operators—or those that execute a control process—including vehicle control, performance of preflight and postflight subsystem checkout, and handling of failure detection and recovery. While Draper Laboratory additionally received a patent on Timeliner in 1998, the software was licensed to Auspice in 1997, for exclusive use in commercial markets.

"We took the baseline version Timeliner and commercialized it as TLX," said Rick Berthold, Draper's former ISS Timeliner lead engineer who cofounded Auspice and remains as its chief technical officer. "The original Timeliner gave us the revolutionary English-like development language and an ultra-reliable runtime environment. We greatly extended the utility of the language, added interconnectivity to networking devices, servers, databases, and application software, and built a limitless distributed execution environment that could process up to millions of simultaneous events."

As a software platform, Auspice TLX uses the English-like language Berthold referred to, called MetaScript,™ to provide rapid development and scalable, real-time execution environments. The role of the TLX MetaScript is to marshal and coordinate the capabilities of the disparate enterprise systems in order to accomplish an activity or "mission." The language was designed in an accessible format so that content experts who do not know how to write computer programs can quickly build TLX applications, as well as anticipate and respond to events in real time. Many of the functions provided by the TLX MetaScript parallel those of a human operator for executing network and

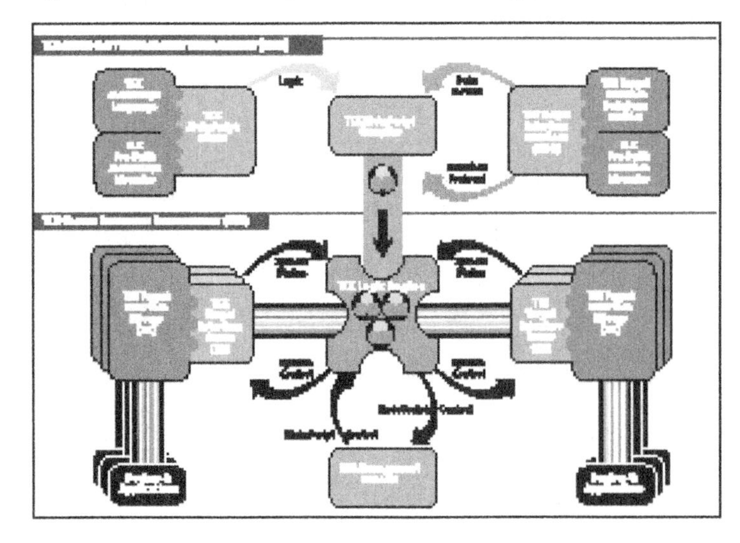

TLX® includes (1) a development environment with natural-language scripting and systems interface definitions for business logic, and (2) an execution environment with execution engines plus device and target interface libraries and servers that can carry out business logic on networked devices, applications, or databases.

business operations. Just like a human operator, the TLX MetaScript provides a level of functional flexibility that cannot be achieved using lower-level language-based software systems. Unlike human operators, though, the automated language reacts rapidly with error-free repetition.

To the advantage of its users, Auspice TLX possesses an enormous library of pre-designed, industry standard interfaces that are ready to plug MetaScript routines or control commands into virtually any hardware device, database, or application program. Just as important is Auspice TLX's ability to execute actions in real time. By instantly and continuously monitoring and correlating events and networks, the technology has total control in ensuring accuracy within business operations. For instance, businesses employing Auspice TLX can expect to see major improvements in efficiencies associated with their billing systems, order entries, and customer service. On the whole, the amalgamation of the MetaScript language, the interface collection, and the real-time monitoring functionality makes the product an optimum end-to-end business solution.

Auspice has so far focused on the broadband, telecommunications, and cable television industries. The company's early success includes an array of "multi-service operator" customers, including Comcast Corporation, Cablevision Systems Corporation, and Insight Communications; and system integration partners, such as Computer Sciences Corporation. With millions of customers subscribed to a growing list of cable-based multi-services like Internet access, telephony services, and video-on-demand, tremendous orchestration is needed to facilitate these operations. When cable shuts down in one particular area, for instance, cable companies receive hundreds to thousands of alerts to notify them of the problem. The disturbance can come from high up in the network management system, or it can be linked to one single household. Even so, it can take hours or sometimes days to pinpoint and troubleshoot the root cause.

The TLX product addresses many pervasive issues that plague service providers' operations. Broadband, telecommunications, and cable companies use it to create applications that provide a real-time, holistic view of their operations—correlating millions of events and integrating service provisioning, customer care, field service operations, and the physical network. The real-time integration and automation feature, for example, will let broadband service providers know exactly when and where they need to send out a service technician—be it to an individual subscriber's home or to a neighborhood or regional node, how to rectify a customer service problem with the touch of a single button, and how to prevent a performance problem from escalating into a full-blow service outage.

When the TLX-based application detects problems through its real-time monitoring and correlating capabilities, a quick fix is on the way. The software not only gives this "holistic view" to in-house network operation centers (NOC) and customer service staff, it can deliver the same information right to a service technician's wireless handheld computer out in the field. Auspice TLX can also provide a geographic information systems link so that a NOC or a dispatched field technician can map out the exact location of a problem's root cause. With features like these, Auspice's customers have documented faster mean-time-to-repair, fewer repair "truck rolls," and higher staff productivity.

In addition to using TLX to custom-build applications for its customers, Auspice has released the OpsLogic™ Broadband Solution Suite, an integrated service provider operations solution powered by Auspice TLX. The product suite consists of five packaged-yet-customizable applications to address the challenges of managing the provisioning, reliability, and repair of broadband, telephony, high-speed data, video-on-demand, and other services.

While the broadband market still remains fertile, Auspice contends that many new markets will open up for Auspice TLX and OpsLogic over the next year. Meanwhile, the company will continue its efforts to bridge the gap between networks, business operations, and individual users in selected markets, just as Timeliner will work to bridge the gap between Earth and space for NASA.

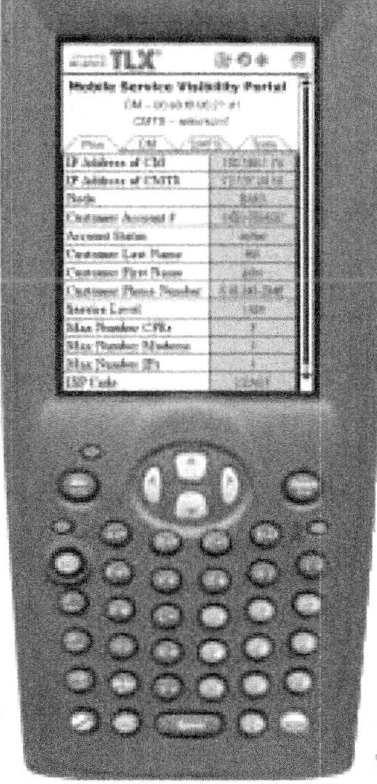

TLX® sends real-time information and interactive tools right to network operations center managers' and customer service representatives' desktops—even to field technicians' wireless handheld computers.

Lending a Helping Hand

Barrett Technology,® Inc., of Cambridge, Massachusetts, received the 2003 Robotic Industries Association's Joseph Engelberger Award for Technology Leadership based on successful commercialization of its novel robotic manipulators. Designed for applications requiring superior adaptability, programmability, and dexterity, Barrett's devices provide state-of-the-art functionality and capability, as well as product integration with existing technology. The cutting-edge robotic manipulators originated through collaboration with NASA, the National Science Foundation, and the U.S. Air Force.

In the 1990s, NASA's Johnson Space Center awarded Barrett four **Small Business Innovation Research** (SBIR) contracts, leading the company to develop the first commercially available cable-driven robots. Today, the company supports two robotic manipulator product lines: the Whole-Arm Manipulation System (WAM™) and its BH8-Series™ hands, both of which received funding through SBIR contracts. During a Phase II SBIR contract with Johnson, Barrett designed the EVA-Retriever WAM arm for NASA's use as an autonomous robot to recover crew or tools outside of the Space Station.

The WAM arm outperforms today's conventional robots through its extraordinary dexterity, transparent dynamics, high bandwidth, zero backlash, and near-zero friction. The device can reach around objects and clasp them, much like a person holding a large item between his or her forearm and upper arm without compromising the use of hands for small items. Conventional robotic arms are strictly limited to the use of hand end-effectors and therefore small payloads. The WAM arm is also distinguished from other arms with its use of gear-free cable drives to manipulate its joints.

Listed in the Millennium Edition of The Guinness Book of World Records (2000) as the world's most advanced robotic arm, the WAM arm closely resembles its human counterpart. The arm consists of a shoulder that operates on a gearless differential mechanism, an upper arm, a gear-free elbow, forearm, and wrist. This arrangement of joints coincides with the human shoulder and elbow, but with much greater range of motion. Like a person's arm, but unlike any industrial robotic arm, the

The WAM™ arm consists of a shoulder that operates on a gearless differential mechanism, an upper arm, a gear-free elbow, forearm, and wrist. This arrangement of joints coincides with the human shoulder and elbow, but with much greater range of motion.

WAM arm is backdriveable, meaning that any contact force along the arm or its hand is immediately felt at the motors, supporting graceful control of interactions with walls, objects, and even people. With a human-scale 3-foot reach, it is so quick that it can grab a major-league fastball, yet so sensitive that it responds to the gentlest touch.

Like the WAM arm, the BH8-262 BarrettHand™ offers many benefits in dexterity. A multi-fingered programmable grasper, the BarrettHand can pick up objects of different sizes, shapes, and orientations. According to the company, integrating this device with any robotic arm is fast and simple, and immediately multiplies the value of any arm requiring flexible automation. The BarrettHand is compact and completely self-contained, weighing only 1.18 kilograms (kg). The newest product in the series is the soon-to-be-released BH8-601 Wraptor,™ a large-capacity 7-kg, three-fingered system featuring enhanced dexterity, a vision-camera mount, and user-accessible sensor support on all finger and palm surfaces. In addition to wrapping its fingers around an object, the device can perform internal grasps by reaching its fingers inside of an object and then spreading them open.

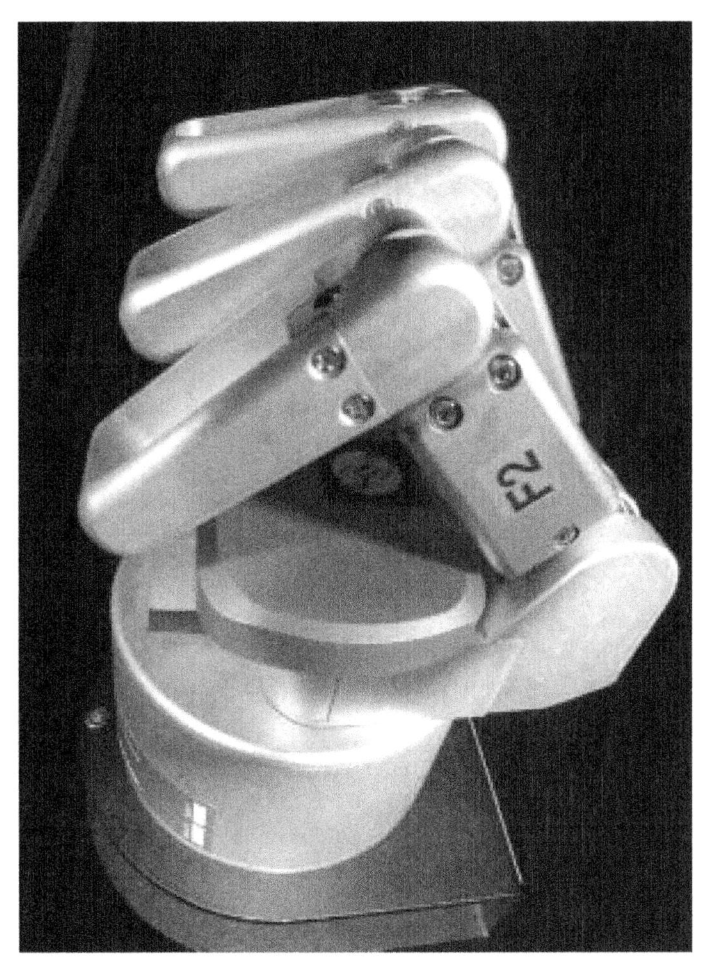

The BarrettHand,™ a multi-fingered programmable grasper, can pick up objects of various shapes and sizes.

The WAM arm, BarrettHand, and Wraptor have commercial applications ranging from human-collaborative medical surgery to emergency response to chemical, biological, and nuclear materials. Barrett is also targeting markets such as physical therapy, rehabilitation, assisted-living aids, metrology, short-run manufacturing, and entertainment.

WAM,™ BH8-Series,™ BarrettHand,™ Wraptor,™ and the Barrett Technology® logomark are trademarks of Barrett Technology, Inc.

Nanoscale Liquid Jets Shape New Line of Business

Just as a pistol shrimp stuns its prey by quickly closing its oversized claw to shoot out a shock-inducing, high-velocity jet of water, NanoMatrix, Inc., is sending shockwaves throughout the nanotechnology world with a revolutionary, small-scale fabrication process that uses powerful liquid jets to cut and shape objects.

Emanuel Barros, a former project engineer at NASA's Ames Research Center, set out to form the Santa Cruz, California-based NanoMatrix firm and materialize the micro/nano cutting process partially inspired by the water-spewing crustacean. Early on in his 6-year NASA career, Barros led the development of re-flown flight hardware for an award-winning Spacelab project called "NeuroLab." This project, the sixteenth and final Spacelab mission, focused on a series of experiments to determine the effects of microgravity on the development of the mammalian nervous system.

In 1999, Barros transitioned into a project supporting the development of International Space Station research hardware, and was considered a nanotechnology expert among many of his peers in the Life Sciences Division at Ames. Fully satisfied with his accomplishments at NASA, Barros departed Ames in 2002 to succeed in nanoscale manufacturing as the chief technical officer and acting chief executive officer of NanoMatrix.

In addition to cutting and shaping, NanoMatrix's proprietary machining services and equipment are capable of performing sub-micron etching, drilling, and welding, all with nanometer precision. "At the scale that we are working, the jets use individual molecules to produce a machining effect," Barros explains. The processes pioneered by NanoMatrix are environmentally friendly, unlike semiconductor chip photolithography, which makes use of toxic chemicals. NanoMatrix employs mechanical methods, so no harsh chemicals are required. The processes are also extremely tolerant of contaminants, so they can be used in environments with less stringent cleanliness controls.

The company's work represents an alternative method for developing and building small-scale electronic, mechanical, and medical devices, among other applications. Until this technology became available, most small-scale fabrication processes were developed using large-scale circuit tools like semiconductor-manufacturing equipment. While these larger-scale tools continue to evolve and enhance the size and yield of two-dimensional circuits, they do not always meet the needs of developers working in other application areas. Leading-edge semiconductor technologies use a limited set of materials and are expensive, according to NanoMatrix. While they produce superb economies of scale, these processes require very large run sizes to be

This laboratory apparatus was used to develop a process in which NanoMatrix's high-velocity liquid jets helped adhere a family of next-generation coatings to materials that would otherwise not adhere very well.

cost effective, the company adds. For this reason, microelectrical mechanical systems (MEMS) or nanoelectrical mechanical systems (NEMS) are traditionally produced using older semiconductor technology. Conversely, NanoMatrix's processes offer flexible run sizes, as large volumes are not required to be cost effective.

With NanoMatrix's liquid jet machining, significant opportunities abound in prototyping and creating smaller MEMS/NEMS devices. Furthermore, the process widens the potential for use of new materials that cannot be shaped or etched using semiconductor photolithography or e-beam procedures. To date, the company's biggest application involves film adhesion. NanoMatrix developed a process for a client that allows its next generation of films to adhere to a glass surface, such as a window or a lens, without affecting the clarity or the amount of light that shines through. The process increases the surface area; other procedures that attempt to broaden surface range will cause the glass to become opaque, and therefore useless, according to Barros.

NanoMatrix is also developing an inexpensive process to prevent glare on windows and lenses. This too, could be practical for computer, cell phone, and personal digital assistant screens, which are often subjected to strong sunlight that can obstruct displayed images and strain users' eyesight.

Farther down the road than the film adhesion process for NanoMatrix—but not too far, says Barros—is the likelihood of a first generation of nanotechnology-type mechanical machines that can be manufactured using a general-purpose rapid prototyping and/or production tool. These tiny machines would incorporate designs that could be developed by NanoMatrix or any third party using the company's tools. Such devices could be used as "sensing dust" for collecting sensory data of all kinds, then transmitting the information to a central computer for processing, which in turn, could benefit future space exploration, crime scene investigations, and field operations for military personnel.

NanoMatrix is currently focusing its efforts on joint development projects with customers and business partners. "We have found that there are a number of companies that cannot effectively build their next-generation products using the limited tool kit that is currently available," Barros notes. The projects include a variety of MEMS, microfluidics, and optics companies that are developing solutions to problems associated with fabricating devices with feature sizes often smaller than the wavelength of visible light. In addition, NanoMatrix is constructing a general-purpose workstation for customers wishing to work independently on development and prototyping projects.

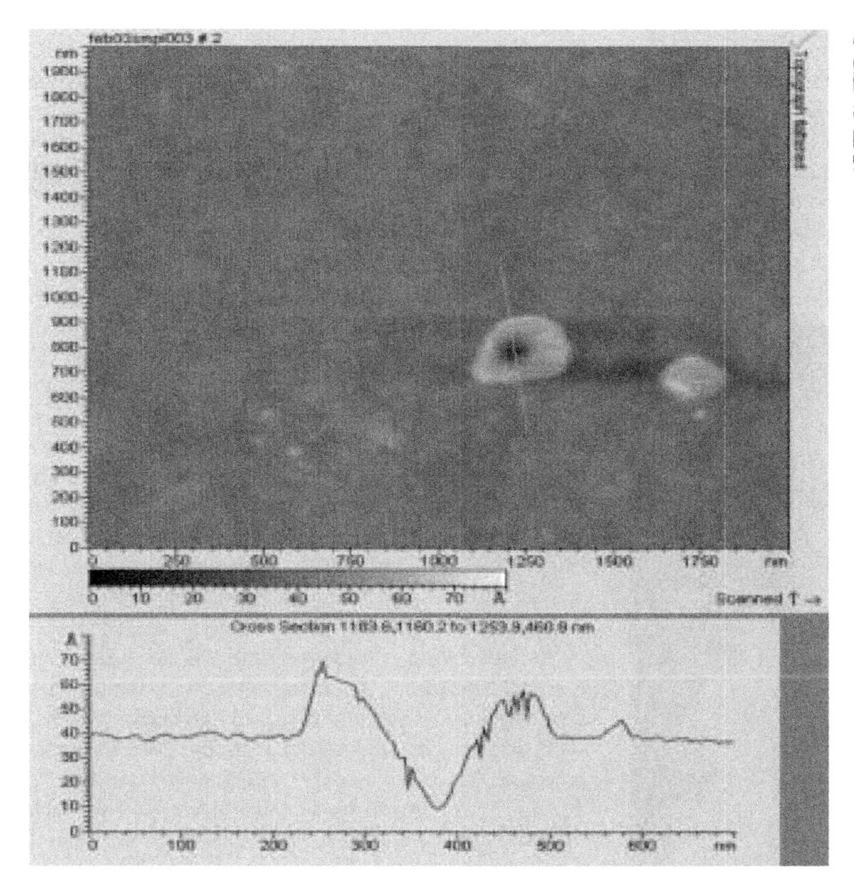

An "Atomic Force Microscope" micrograph shows a nanometer-drilled hole that increases surface area for film adhesion. The line graph represents a profile of the hole, along the bisecting green line.

Solutions for Hot Situations

From the company that brought the world an integral heating and cooling food service system after originally developing it for NASA's Apollo Program, comes yet another orbital offshoot: a product that can be as thin as paper and as strong as steel.

Nextel™ Ceramic Textiles and Composites from 3M Company offer space-age protection and innovative solutions for "hot situations," ranging from NASA to NASCAR. With superior thermal protection, Nextel fabrics, tape, and sleevings outperform other high-temperature textiles such as aramids, carbon, glass, and quartz, permitting engineers and manufacturers to handle applications up to 2,500 °F (1,371 °C). The stiffness and strength of Nextel Continuous Ceramic Fibers make them a great match for improving the rigidity of aluminum in metal matrix composites. Moreover, the fibers demonstrate low shrinkage at operating temperatures, which allow for the manufacturing of a dimensionally stable product. These novel fibers also offer excellent chemical resistance, low thermal conductivity, thermal shock resistance, low porosity, and unique electrical properties.

The origins of Nextel Ceramic Textiles and Composites reach all the way back to the early days of the Space Shuttle Program, when NASA scientists were tasked with improving high-temperature tiles and textiles to withstand the intense heat and pressures of reentry. Through research and testing, the ceramic fibers that are the framework for St. Paul, Minnesota-based 3M's Nextel technology proved to be suitable for use in the composition of the Shuttle's underbelly tiles. As the Space Shuttle Program progressed, the fibers were also woven into sleeving and fabric for use in gap fillers, flexible insulation blankets, heat shields, gaskets, and seals.

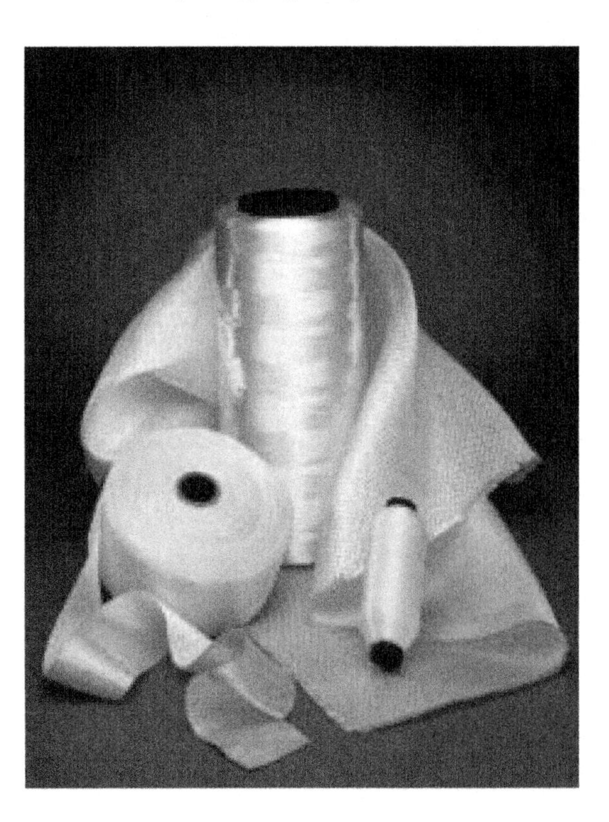

More recently, extensive testing by NASA researchers at the Marshall Space Flight Center and the Johnson Space Center demonstrated Nextel's value as a vital component in the development of a strong, lightweight meteoroid/space debris shield known as a "Stuffed Whipple Shield." This technology, which also encompasses the nylon-like Kevlar® polymer from DuPont, is currently being used to safeguard the International Space Station and the RADARSAT spacecraft from inevitable contact with space debris. The results from the hypervelocity impact testing at Marshall and Johnson showed that Nextel Ceramic Textiles and Composites improve shield performance, when compared to aluminum, because they are "better at shocking projectile fragments," and sustain far less damage after an impact with the debris.

3M and NASA's contribution to improving upon conventional aluminum shielding methods is playing a major role in protecting vehicles, payloads, and crew members by fending off the debris, which continues to generate over time. According to recently published reports consistent with information offered by Johnson's Space Science Enterprise, space launches occurring over the last 40 years have led to more than 23,000 observable objects larger than 10 centimeters, 7,500 of which are still in orbit. These reports also show that in 2000, there were between 70,000 and 120,000 "on-orbit" debris fragments (not attributed to launches, satellites, or mission-related objects) larger than 1 centimeter floating in space.

At the same time 3M was meeting NASA's needs for the development of lightweight textile materials, the company was introducing the Nextel technology to private industry. In commercial aviation, Nextel Flame Stopping Dot Paper provides superior performance for fuselage burn-through protection. Private aircraft, such as the team plane for the National Basketball Association's Miami Heat, also rely on the flame-stopping technology.

With superior thermal protection, the Nextel™ family of products outperforms other high-temperature textiles such as aramids, carbon, glass, and quartz.

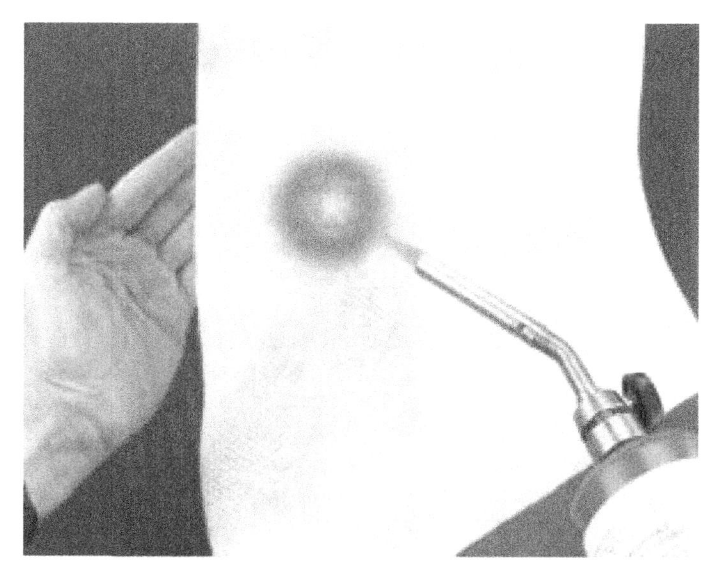

Nextel™ Flame Stopping Dot Paper resists flame penetration for greater than 4 minutes at a temperature of 2,000 °F.

Despite its paper-thin characteristics, Nextel Dot Paper withstands fire without burning or shrinking. The unique "dots" help to maintain the paper's integrity and flexibility. Testing at the Federal Aviation Administration's William J. Hughes Technical Center shows that the Flame Stopping Dot Paper resists flame penetration for greater than 4 minutes at a temperature of 2,000 °F (1,093 °C). In addition to fuselage burn-through applications, Nextel Dot Paper can be used for added protection in galleys, cockpits, cargo bays, firewalls, fire doors, ducting, insulators, gaskets, seals, and fire-resistant storage.

The petrochemical industry is using flexible tube seals fabricated from Nextel Ceramic Textiles and Composites to save energy and improve control over high-temperature chemical processes. For example, the M.W. Kellogg Company, of Houston, Texas, adopted Nextel ceramic tape in the construction of a circulating-bed transport reactor. By choosing Nextel materials, M.W. Kellogg now offers refineries a cleaner form of gas production that creates less pollution and hazardous waste than previous technologies.

When used in industrial furnaces, Nextel woven fabrics can also serve as thermal barriers to separate different temperature zones while preventing particulate shedding. Nextel ceramic sleevings modeled after door gaskets and seals developed for the Space Shuttle are used to seal doors and other access panels on the furnaces.

The Nextel product also helps NASCAR teams reduce the heat transfer from engine, exhaust, and track surface into the driver's compartment. Race car drivers routinely face ambient cockpit temperatures of 115 °F (46 °C) or more and floorboard temperatures hot enough to boil water. Lightweight Nextel Thermal Barriers effectively block heat before it gets into the vehicle. Tested during the 1999 season on NASCAR Winston Cup and Craftsman Truck Series vehicles, the thermal barriers are now used on Winston Cup, Busch Grand National, and Craftsman Truck vehicles to reduce driver compartment temperature by more than 70 °F (21 °C).

Nextel™ is a trademark of 3M Company.
Kevlar® is a registered trademark of E.I. du Pont de Nemours and Company.

Extensive testing by NASA researchers at the Marshall Space Flight Center and the Johnson Space Center demonstrated Nextel's™ value as a vital component in the development of a strong, lightweight meteoroid/space debris shield known as a "Stuffed Whipple Shield."
Images courtesy of 3M Company (www.3m.com).

Prompt and Precise Prototyping

For Sanders Design International, Inc., of Wilton, New Hampshire, every passing second between the concept and realization of a product is essential to succeed in the rapid prototyping industry—where amongst heavy competition, faster time-to-market means more business.

To separate itself from its rivals, Sanders Design aligned with NASA's Marshall Space Flight Center to develop what it considers to be the most accurate rapid prototyping machine for fabrication of extremely precise tooling prototypes. The company's Rapid ToolMaker™ System has revolutionized production of high quality, small-to-medium sized prototype patterns and tooling molds with an exactness that surpasses that of computer-numerically-controlled (CNC) machining devices.

Created with funding and support from Marshall under a **Small Business Innovation Research (SBIR)** contract, the Rapid ToolMaker is a dual-use technology with applications in both commercial and military aerospace fields. The advanced technology provides cost savings in the design and manufacturing of automotive, electronic, and medical parts, as well as in other areas of consumer interest, such as jewelry and toys. For aerospace applications, the Rapid ToolMaker enables fabrication of high-quality turbine and compressor blades for jet engines on unmanned air vehicles, aircraft, and missiles.

The Rapid ToolMaker can generate numerous compound surface features not possible with CNC machines. For example, it can easily produce complex turbine-bladed disk (blisk) patterns, comprised of the hub, blades, and outer retaining rims, as a single unit. It can also construct miniature air passages as small as 75 microns to facilitate air cooling of critical airfoil surfaces, with entrance and exit ports near or on the leading and trailing edges of the blades, respectively. These patterns undergo direct investment casting to yield composite metal turbine blades. Patterns fabricated on the Rapid ToolMaker enable future availability of replacement patterns for manufacturing operations, as well as long-term logistic and maintenance support.

Designed for unattended operation, the Rapid ToolMaker is a freestanding unit that supports a computer-aided design (CAD) workstation environment. It sequentially builds one layer at a time from a stereolithography file or a Hewlett-Packard Graphic Language file; Microsoft® Windows®-based graphical user interface software provides direct use and editing of various other CAD file formats.

The system's ink-jet deposition process for layered fabrication of 3-dimensional prototypes improves accuracy and surface finish quality by a factor of 10, compared with other rapid prototyping technologies, according to Sanders Design. A combination of build and support material is deposited as low-viscosity "ink" by dual ink-jets, which glide over a build platform on a precision, computer-driven carriage. This technique ensures layer uniformity, registration, and exact replication of a CAD design file. A solvent then removes the sacrificial support material without additional post-processing. The completed prototype pattern is dimensionally accurate with a smooth surface finish, and is used directly for metal casting or mold forming.

The technology has several automated features that ensure proper model construction and quality. Print head operation is checked under program control by moving the print head assembly to a validation station to monitor proper material flow. Heated material reservoirs hold the material for 100 hours of continuous operation, and replenishment does not interrupt a job in process. Additionally, a controller monitors and directs all operations under software control for the entire length of the fabrication.

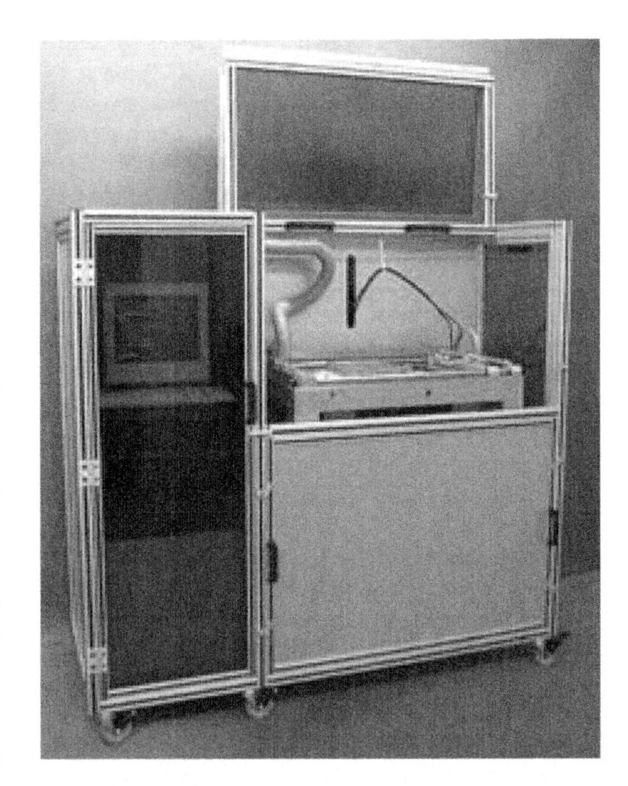

According to Sanders Design International, Inc., the Rapid ToolMaker™ is the most accurate free-form fabrication system in the world. It enables rapid prototyping of precision tooling patterns for aerospace, medical, electronic, automotive, and consumer products direct from computer-aided design files using layered fabrication techniques and ink-jet deposition technology.

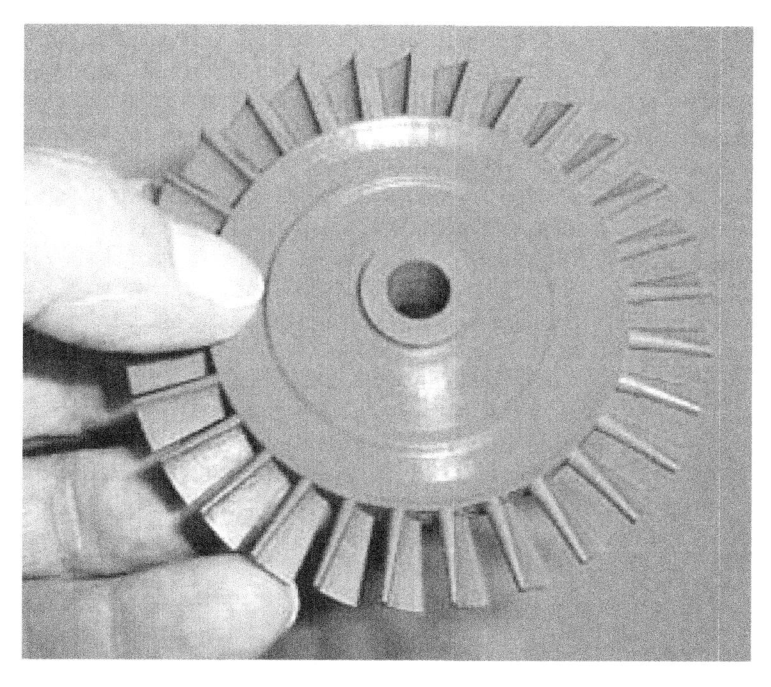

Bladed disks (blisks) and turbine blades for small- to full-scale jet engines are produced by the Rapid ToolMaker™ system with unsurpassed fidelity, precision, accuracy, detail, surface finish, and repeatability for missile and commercial airliner applications. Using a sacrificial and dissolvable support material, the RTM can repeatably produce any blisk design which cannot be fabricated using traditional prototyping, computer-numerically-controlled devices, or other special tooling methods.

The Rapid ToolMaker's accolades include being named the grand winner at the 1997 NASA-sponsored "Technology 2007" Conference in Boston, Massachusetts, and winning the SBIR "Technology of the Year" Award from the Technology Utilization Foundation in the same year.

Sanders Design is currently participating in Phase II of a separate SBIR contract with NASA's Langley Research Center to enhance the Rapid ToolMaker. The goal of this partnership is to fabricate ceramic models with self-contained sensor conductors simultaneously in a single unit, for wind tunnel testing that could potentially lead to advanced Space Shuttle design. Direct fabrication of ceramic parts has multiple benefits for producing low-cost precision devices that can withstand extremely harsh environments encountered in automotive and aerospace applications.

The Rapid ToolMaker's™ ability to position ink-jet droplets to within ±5 microns enables fine-featured jewelry and other prototypes to be fabricated in a flawless manner. This diamond-studded earring prototype is shown expanded six times through a jeweler's eye loupe.

Multiplying Electrons With Diamond

As researchers in the Space Communications Division of NASA's Glenn Research Center in 1992, Dr. Gerald Mearini, Dr. Isay Krainsky, and Dr. James Dayton made a secondary electron emission discovery that became the foundation for Mearini's company, GENVAC AeroSpace Corporation. Even after Mearini departed Glenn, then known as Lewis Research Center, his contact with NASA remained strong as he was awarded **Small Business Innovation Research** (SBIR) contracts to further develop his work.

Mearini's work for NASA began with the investigation of diamond as a material for the suppression of secondary electron emissions. The results of his research were the opposite of what was expected—diamond proved to be an excellent emitter rather than absorber. Mearini, Krainsky, and Dayton discovered that laboratory-grown diamond films can produce up to 45 electrons from a single incident electron. Having built an electron multiplier prototype at NASA, Mearini decided to start his own company to develop diamond structures usable in electron beam devices.

Mearini proposed and received Phase I SBIR funding from Glenn to continue his work to prove diamond's potential as a device material. Since then, his Cleveland, Ohio-based company has improved secondary electron emissions from laboratory-grown diamond to produce greater than 60 electrons from a single impinging electron. Diamond produces 15 times more electrons than other materials used to amplify electrons, giving diamond the highest known secondary electron emission coefficient.

Companies making electron beam devices use GENVAC's diamond product. GENVAC produces chemical vapor deposited diamond as a dynode for electronic and electro-optical device components. Diamond dynodes allow manufacturers to make devices considerably smaller, in addition to increasing their efficiency and reliability. Applications include electron beam amplification devices such as night vision goggles and specialized sensors, field emission cathodes for flat panel displays, and heat spreaders for thermal management.

In another field, a leading manufacturer and supplier of specialized electron tubes and electro-optic products applied GENVAC's diamond to its dynodes for photo multiplier tubes (PMT) with medical applications. Beginning with the production of diamond-based PMTs in late 2000, the technology is leading the way for smaller, more efficient medical devices.

Dayton, who was Glenn's Electron Beam Technology Branch Chief, retired from NASA in 1998 to become Director of Technology at Hughes (now Boeing) Electron Dynamic Devices, in Torrance, California. After 3 years of commuting between Los Angeles and Cleveland, he resigned from Boeing and joined GENVAC as Director of Technology in November 2001.

GENVAC continues to work with NASA through the SBIR program. The company was recently awarded a Phase I SBIR contract from NASA's Jet Propulsion Laboratory to develop a diamond-based sub millimeter backward wave oscillator. The company has also been selected by Glenn to develop a diamond-based 60-Gigahertz miniature Traveling Wave Tube.

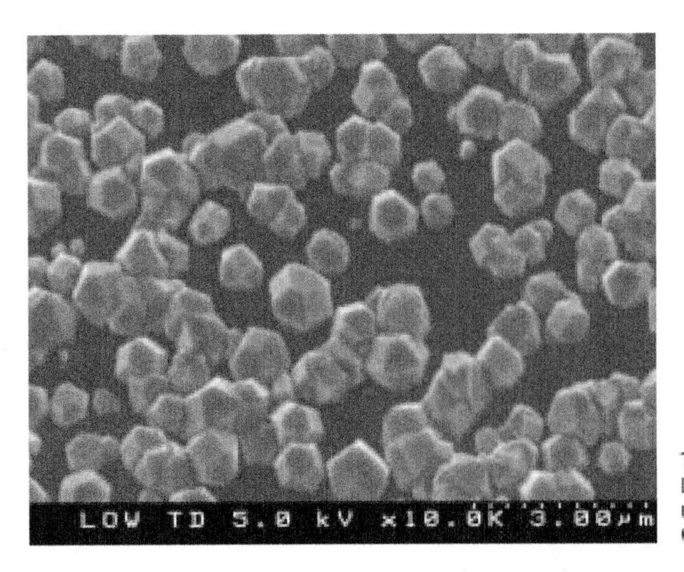

This image shows a diamond field emitter. Laboratory-grown diamond films can produce up to 45 electrons from a single incident electron, making them an excellent emitter.

Conducting the Heat

Heat conduction plays an important role in the efficiency and life span of electronic components. To keep electronic components running efficiently and at a proper temperature, thermal management systems transfer heat generated from the components to thermal surfaces such as heat sinks, heat pipes, radiators, or heat spreaders. Thermal surfaces absorb the heat from the electrical components and dissipate it into the environment, preventing overheating.

To ensure the best contact between electrical components and thermal surfaces, thermal interface materials are applied. In addition to having high conductivity, ideal thermal interface materials should be compliant to conform to the components, increasing the surface contact. While many different types of interface materials exist for varying purposes, Energy Science Laboratories, Inc. (ESLI), of San Diego, California, proposed using carbon velvets as thermal interface materials for general aerospace and electronics applications.

NASA's Johnson Space Center granted ESLI a **Small Business Innovation Research (SBIR)** contract to develop thermal interface materials that are lightweight and compliant, and demonstrate high thermal conductance even for nonflat surfaces. Through Phase II SBIR work, ESLI created Vel-Therm® for the commercial market. Vel-Therm is a soft, carbon fiber velvet consisting of numerous high thermal conductivity carbon fibers anchored in a thin layer of adhesive. The velvets are fabricated by precision cutting continuous carbon fiber tows and electrostatically "flocking" the fibers into uncured adhesive, using proprietary techniques.

Johnson granted ESLI a Phase III SBIR contract to evaluate its thermal interface material for applications on the International Space Station (ISS). Avionics and batteries mounted externally on the ISS must be easily replaceable by astronauts or robotics. To serve ISS needs, the components are packaged as orbital replaceable units (ORUs). These ORUs require thermal contact with the ISS thermal control system to keep the component temperatures within required limits. The current design of the ORU thermal interface consists of interleaving black-anodized aluminum fins which radiatively exchange heat. Adding ESLI Vel-Therm to the fins of the ORU improves thermal contact by replacing radiative with conductive heat exchange. The removable ORU fins slide into place over the fins of the mating heat sink mounted on the ISS.

Vel-Therm has commercial applications in the aerospace and electronics industries. It particularly benefits applications with large or uneven gaps or sliding interfaces. Each fiber provides a thermal path from the electronic component to the thermal surface. In addition to having high conductivity, the material makes intimate contact with interfacing surfaces because of its high compliance. Each fiber bends independently, allowing the fibers to come into contact with both surfaces, even when the surfaces are not parallel, flat, or smooth. A minimal amount of pressure is required for intimate contact, precluding the need for heavy bolts or clamping mechanisms, and eliminating the necessity of flat, smooth mating surfaces.

This finned interface prototype of an orbital replaceable unit for the International Space Station has Vel-Therm® applied to every other fin slot.

A Closer Look at Quality Control

Spectrometers, which are durable, lightweight, and compact instruments, are a requirement for NASA deep space science missions, especially as NASA strives to conduct these missions with smaller spacecraft. NASA's Jet Propulsion Laboratory (JPL) awarded the Brimrose Corporation of America a **Small Business Innovation Research** (SBIR) contract to develop a compact, rugged, near-infrared spectrometer for possible future missions.

Spectrometers are of particular importance on NASA missions because they help scientists to identify the make-up of a planet's surface and analyze the molecules in the atmosphere. Minerals and molecules emit light of various colors. The light, identified as spectra, is difficult to see, and spectrometers, which are essentially special cameras that collect the separate colors of light in an object, allow scientists to identify the different materials. For example, spectrometers can help scientists determine whether soil was created from lava flows or from meteorites.

Brimrose's work for JPL led to a spectrometer that created a significant breakthrough in optical spectrometry. As a result, the Baltimore, Maryland-based company developed the commercially available Luminar 5030 Mini Spectrometer. This innovation is a hand-held analytical instrument utilizing Brimrose's high performance Acousto-Optic Tunable Filter-Near Infrared technology. The product delivers rapid, multi-component analysis for a wide variety of commercial applications, making it a powerful new tool for quality control and process troubleshooting.

The Luminar 5030 can be applied to materials such as powders, polymer pellets, paper goods, fruits and grains, liquids, and vines. The tool performs measurements for moisture, active ingredients, coating weight, and blend assays, as well as for fat, sugar, and protein. The Luminar 5030 can be used for monitoring and control processes in the chemical and petrochemical industry, the inspection of tablets and powder mix control in the pharmaceutical industry, and composition monitoring in the food and beverage industry. For example, Brimrose markets the Luminar 5030 to wineries, enabling vintners to analyze grapes prior to producing wine. The tool's versatility extends from the laboratory to the production floor and field, providing nondestructive and contact/noncontact testing and inspection.

In the pharmaceutical industry, Brimrose's optical spectrometers have passed Good Manufacturing Practice requirements for hardware and software, granting them Food and Drug Administration certification for drug manufacturing by both AstraZeneca, of Wilmington, Delaware, and Pfizer, Inc., of New York, New York. AstraZeneca has chosen Brimrose systems as the only process control spectrometer to be used for manufacturing its pharmaceutical drugs.

Brimrose is continuing to partner with NASA through the SBIR program. The company is in the process of developing a space-qualified optical spectrometer intended for use onboard the land rover vehicles for investigating the surface and subsurface of Mars.

The Luminar 5030 Mini Spectrometer enables vintners to analyze grapes prior to producing wine.

Remote Transmission at High Speed

NASA's need for a more accurate way to collect data from propulsion tests provided Omni Technologies, of New Orleans, Louisiana, an opportunity to codevelop and market the solution. The FOTR-125, a redundant fiber-optic transceiver for the remote transfer of high-speed digital data, is benefiting NASA while also making a commercial impact.

Stennis Space Center, NASA's field center for rocket propulsion testing, has always faced the dilemma of collecting accurate data from the inherent hostile environment of rocket engine tests. Two options were available for data collection: transmit the analog signals from a test stand to a safe location over long copper cables, risking the signals becoming corrupt through the pickup of electromagnetic noise from the environment; or tape record the analog signal on the test stand and transport the tapes to a safe location for processing—a process that would require extra time to digitize the data. The ideal solution was to digitize the analog signal on the test stand and then immediately transmit this digital data for recording in a safe location. To arrive at this solution, various technological and bandwidth constraints had to be overcome.

Omni and NASA Test Operations at Stennis entered a Dual-Use Agreement to develop the FOTR-125, a 125 megabit-per-second fiber-optic transceiver that allows accurate digital recordings over a great distance. The transceiver's fiber-optic link can be as long as 25 kilometers. This makes it much longer than the standard coaxial link, which can be no longer than 50 meters. According to Joey Kirkpatrick, a NASA engineer and codeveloper of the device, "Stennis needed a method to extend this transmission distance, and converting the existing copper communications interface to fiber optic was the obvious solution." The FOTR-125 utilizes laser diode transmitter modules and integrated receivers for the optical interface. Two transmitters and two receivers are employed at each end of the link with automatic or manual switchover to maximize the reliability of the communications link.

NASA uses the transceiver in Stennis' High-Speed Data Acquisition System (HSDAS). The HSDAS consists of several identical systems installed on the Center's test stands to process all high-speed data related to its propulsion test programs. These transceivers allow the recorder and HSDAS controls to be located in the Test Control Center in a remote location while the digitizer is located on the test stand. Using the transceivers in the HSDAS provides a more reliable, capable, and flexible high-frequency data system than previously achievable.

Several versions of the FOTR-125 have been developed allowing additional flexibility. Both redundant and nonredundant versions are available, and the unit may be packaged either as a stand-alone unit or as a rack-mountable chassis unit supporting up to 10 transceivers offering multiple channel capability. The rack-mount chassis also provides for a direct-current battery power supply as a back-up power source. In addition, a daughter card mounting on the FOTR-125 printed-circuit board has been developed, which provides direct access to the Transparent Asynchronous Transceiver Interface (TAXI)-encoded data stream. The FOTR TAXI interface is designed to work with Integrated System Consultants' Direct to Disk system or can be monitored directly with a differential parallel interface.

The Technology Transfer Office at Stennis granted Omni an exclusive license to commercialize the FOTR-125. It is normally packaged as a stand-alone transceiver with built-in power supplies, although other form factors can be accommodated. Omni markets the device to facilities that perform extremely hazardous testing, such as explosives, ordnance, nuclear, rocket engines, and some combustion turbine engines. For government applications, the use of the transceivers at Stennis will continue to grow as incremental upgrades to the HSDAS take place.

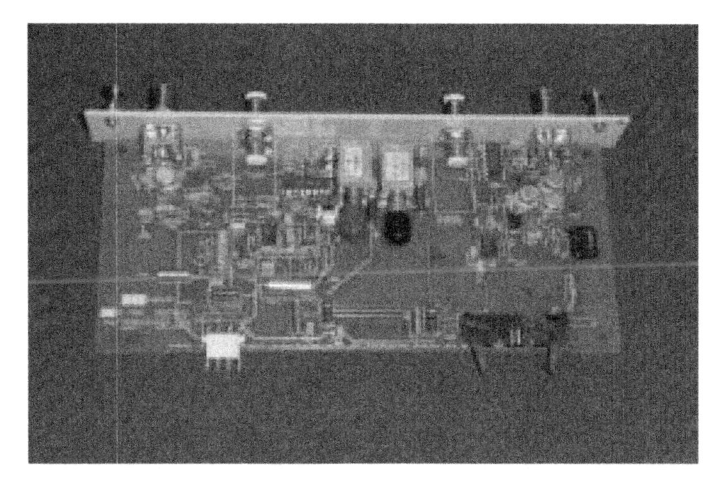

Omni Technologies' FOTR-125 is a redundant fiber-optic transceiver that remotely transfers high-speed digital data.

Going Places No Infrared Temperature Devices Have Gone Before

A leading manufacturer of noninvasive, infrared temperature sensors, Exergen Corporation prides itself on delivering a highly accurate and reliable product to exceed customers' expectations. The Watertown, Massachusetts-based company's forte is designing custom infrared thermocouples (IRt/c's) for professional and consumer use in a wide variety of industrial and medical applications. In 2002, NASA's Glenn Research Center approached Exergen with a simple request to modify the company's product line, reducing the size of its sensors. The outcome led to the development of a family of top-notch IRt/c devices that are pushing production line performance to record highs.

Glenn was seeking an infrared temperature sensor with an extremely small head for a joint flywheel research project with the University of Texas' Center for Electromechanics. Already having created the smallest infrared temperature device available in one package (.5 inches in diameter and 1.45 inches in length), Exergen was up for the challenge. The company claimed that making a shorter unit was easy; however, this was not the area Glenn was hoping to address. Engineers at the NASA research center required a smaller diameter of a quarter-inch for their project, half the size of Exergen's existing device. With funding and support from Glenn, Exergen prevailed in meeting the NASA specifications, and the patented IRt/c™ technology was born.

Exergen's IRt/c is a self-powered sensor that matches a thermocouple within specified temperature ranges and provides a predictable and repeatable signal outside of this specified range. Possessing an extremely fast time constant, the infrared technology allows users to measure product temperature without touching the product. The IRt/c uses a device called a thermopile to measure temperature and generate current. Traditionally, these devices are not available in a size that would be compatible with the Exergen IRt/c, based on NASA's quarter-inch specifications. After going through five circuit designs to find a thermopile that would suit the IRt/c design and match the signal needed for output, Exergen maintains that it developed a model that totaled just 20 percent of the volume of the previous smallest detector in the world.

Following completion of the project with Glenn, Exergen continued development of the IRt/c for other customers, spinning off a new product line called the micro IRt/c. This latest development has broadened applications for industries that previously could not use infrared thermometers due to size constraints. The first commercial use of the micro IRt/c involved an original equipment manufacturer that makes laminating machinery consisting of heated rollers in very tight spots. Accurate temperature measurement for this application requires close proximity to the heated rollers. With the micro IRt/c's 50-millisecond time constant, the manufacturer is able to gain closer access to the intended temperature targets for exact readings, thereby increasing productivity and staying ahead of competition.

Image courtesy of Ola Nordell MotorSports.

Exergen Corporation's noncontact IRt/c™ sensor (middle) measures tire temperature for the Ola Nordell Racing Team's top dragster. An accompanying laser is used to measure ride height and traction.

The IRt/c™ technology is mounted on a remote station monitoring snow cover on the famous Matterhorn in the Swiss Alps.

All Exergen IRt/c sensors come hermetically sealed in a range of sizes and configurations, depending upon the application. They are designed for years of trouble-free operation in the toughest environments. In fact, IRt/c sensors are so rugged, professional auto racing teams use them to measure critical temperature variables during competition. Such critical variables include tire temperature, which directly affects tire adhesion and wear characteristics, and provides valuable data on the set-up and performance of the suspension. For example, excessive loading of a tire caused by out-of-tune suspension will cause it to become considerably warmer than the vehicle's other tires. Exergen IRt/c's were displayed in action at the 2002 Race-A-Rama show in Springfield, Massachusetts. Dittman & Greer, Inc., an Exergen distributor from Middletown, Connecticut, used the technology to measure torque converter, tire, and track temperature throughout the event.

In a separate application, the infrared temperature sensor is being utilized for avalanche warnings in Switzerland. The IRt/c is mounted about 5 meters above the ground to measure the snow cover throughout the mountainous regions of the country. The sensor is part of a larger "snow-station" that records ambient temperature, solar radiation, snow height, snow drift, wind flow, wetness, and ground and air motion.

With IRt/c, Exergen offers system solutions and technical application support for any thermal process requiring precise temperature measurement or control. More than 300 models of noncontact infrared sensors are available to provide accurate, reliable, and cost-effective temperature measurements and control for the most demanding applications.

IRt/c™ is a trademark of Exergen Corporation.

New Sensor Gaining Interest on Industry Radar Screen

While radar is typically used to track large objects that are relatively far away, an Atlanta, Georgia-based start-up company is using the technology in a counter-intuitive way to track very small changes in displacement at close proximity.

Radatec, Inc., a designer, manufacturer, and implementer of sensor systems for monitoring combustion-zone components in turbine engines, was formed in 2001 to commercialize the patented radio frequency vibrometer technology from the Georgia Institute of Technology (Georgia Tech). In readying the technology for the commercial market, Radatec received assistance from NASA's Dryden Flight Research Center in the form of Phase I and Phase II Small Business Innovation Research (SBIR) grants totaling over $650,000. The company's Phase II submission was ranked the number one SBIR proposal out of Dryden for 2002.

Radatec, Inc., Cofounder and President Scott Billington studies his company's noncontact microwave sensor. The patented technology was commercialized by Radatec to monitor complex heavy machinery used on aircraft and in power plants.

The SBIR contracts helped fund further development of Radatec's proprietary, noncontact microwave sensor technology for monitoring complex heavy machinery used on aircraft and in power plants, such as gas and steam turbines. The ensuing hardware/software system can warn of impending problems before they become dangerous. According to Scott Billington, co-founder of Radatec and former research faculty at Georgia Tech's Manufacturing Research Center, the sensor gives the operators of heavy machinery the capability to measure critical pieces of equipment in hostile environments where conventional sensor technology cannot operate. "Our radar-based system will enable operators to more effectively schedule costly maintenance and to run equipment longer. This will help reduce the cost of operating and maintaining these systems," Billington elaborated.

Since vibration measurements are the most important and informative indicator for accessing the health of rotating machinery, Radatec's sensor systems have found their widest application in protecting the investment of large, critical rotating equipment. According to the company, there are currently tens-of-thousands of turbines in use for military and commercial aircraft, representing billions of dollars of investment. Significant dollar amounts are spent maintaining the machinery, much of which is consumed in preventative maintenance, inspection, and replacing questionable parts. In power plants, more than 6,000 steam and gas turbines supply America's power grid alone. With downtime costs of $8,000 to $10,000 per hour and maintenance costs that take up 15 percent of the total operating budget, this sector represents an attractive opportunity for Radatec.

The invention has several key capabilities that provide an advantage over existing sensor technologies. It provides very precise data regarding component health while a turbine is in full-speed operation. Previously, operators were forced to remove a turbine from service, dismantle it, and inspect it to determine component health. Moreover, the new sensor measures component deformations with 5-micron resolution, withstands temperatures exceeding 2,500 °F, is unaffected by interference that disrupts other sensors, and accurately assesses vibrations deep within an engine that were previously immeasurable.

In January 2003, Radatec proved just how advantageous its new sensor is. The company installed a sensor system at the Navy's Patuxent River Naval Propulsion Laboratory in Patuxent River, Maryland, as part of a 6-month, head-to-head test of systems to measure the health of a turbine disk that holds turbine blades in place. The competing systems, installed by the Defense Advanced Research Projects Agency (DARPA), consisted of capacitive, laser, and eddy current sensors. The turbine disk that was subjected to the test was pre-stressed with defects and in operation for over 800 hours at 1,200 °F, and at speeds ranging from 1,500 rpm to 18,000 rpm.

Radatec's system performed favorably against the competing sensor systems in terms of data fidelity and probe survivability. Due to its high bandwidth capability, the Radatec system was able to profile the blades with over 2,000 data points per blade, even at the maximum speed of 18,000 rpm. These profiles enabled Radatec to accurately monitor disk failures, based on blade geometry changes over the course of the test.

From a reliability standpoint, Radatec was the only system to complete the test with original equipment. All competing systems required replacing components that failed due to exposure to the high temperature (one system required 14 sets of probes). Toward the end of the test, the disk began to fragment, impacting all of the sensors. Radatec's sensor absorbed the impact and continued to operate, while the other sensors were completely destroyed. Sustained functioning was due to the fundamental nature of Radatec's waveguide probe, which can be constructed of very durable materials and requires no sensitive components for operation. DARPA has since contracted Radatec to install its system for upcoming tests located at the Patuxent laboratory and another major turbine site.

A separate testing initiative demonstrated that the Radatec microwave sensor system would also benefit the automotive industry, when used as a radar technique for active suspension control. The purpose of a vehicle's suspension system is to isolate the shock and vibration caused by road and terrain irregularities from the main body of the vehicle, yet still provide the necessary ground contact to allow the vehicle to stay in control. Components in such suspension systems have traditionally been passive springs and dampers. Active suspension systems, on the other hand, are made up of components that can alter their properties or add energy to the suspension. They were developed to increase performance and handling of the vehicle and the durability of the components, improve rider comfort, and allow for faster speeds over irregular terrains.

Radatec's radar vibration sensing technology is also well-suited for noncontact measurement of road/terrain surfaces immediately ahead of a moving vehicle. This method of sensing is unaffected by dust, debris, rain, or other obscurants, and can be mounted on a vehicle behind a radio-transparent shield, out of harm's way.

In potential future military applications, the microwave sensor could improve the accuracy of fire-on-the-move when vehicles with moving weapons platforms are faced with rough terrain. When added to such a vehicle, like a tank, the sensor technology would provide information to predict the terrain about to be encountered and allow actions to be taken before the event, rather than after.

Radatec is a part of the Georgia Tech's VentureLab program, which assists faculty members in bringing university-based innovations to the commercial marketplace. VentureLab provided early assistance to Billington and Cofounder Jon Geisheimer, who was a faculty member at the Georgia Tech Research Institute when the company was established. During this time, the founders designed and built a laboratory prototype, which was the basis for their application to the NASA SBIR program. A successful Phase I project led to follow-on Phase II funding, and both founders agree that the NASA SBIR awards "provided a true heartbeat for Radatec as an operating company."

Radatec's sensor provides very precise data regarding component health while a turbine is in full-speed operation.

Sensing the Tension

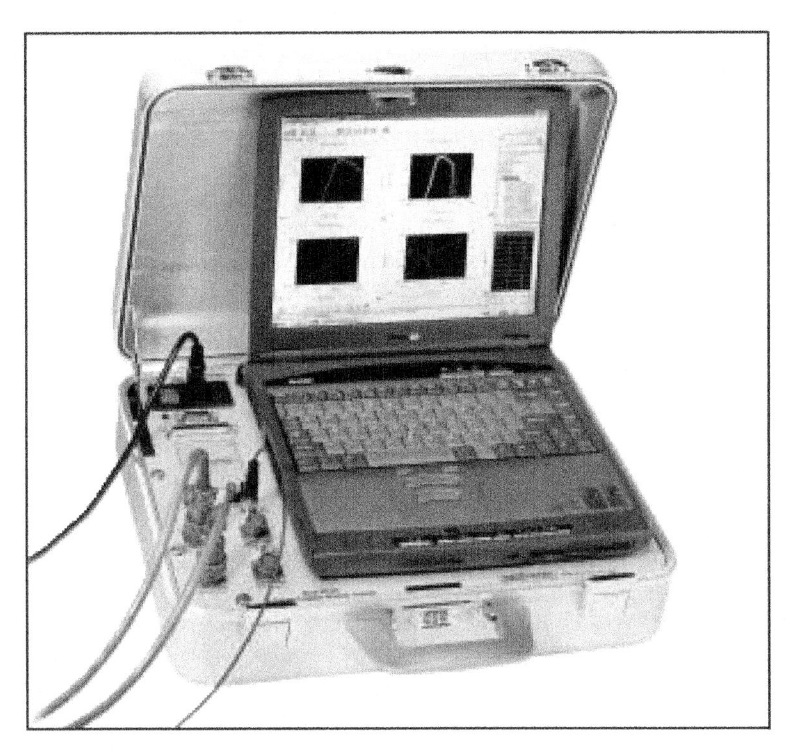

The MC900 Transient Recorder Analyzer is available in a portable laptop model (shown here encased in aluminum), or a desktop model made from high-impact plastic. Standard specifications include four strain gauge input channels, four high-level analog input channels, and four digital input and output channels, among other features.

Spanning over 4 decades, NASA's bolt tension monitoring technology has benefited automakers, airplane builders, and other major manufacturers that rely on the devices to evaluate the performance of computerized torque wrenches and other assembly line mechanisms. In recent years, the advancement of ultrasonic sensors has drastically eased this process for users, ensuring that proper tension and torque are being applied to bolts and fasteners, with less time needed for data analysis.

Langley Research Center's Nondestructive Evaluation Branch is one of the latest NASA programs to incorporate ultrasonic sensors within a bolt tension measurement instrument. As a multi-disciplined research group focused on spacecraft and aerospace transportation safety, one of the branch's many commitments includes transferring problem solutions to industry. In 1998, the branch carried out this obligation in a licensing agreement with Micro Control, Inc., of West Bloomfield, Michigan. Micro Control, an automotive inspection company, obtained the licenses to two Langley patents to provide an improved-but-inexpensive means of ultrasonic tension measurement.

Prior to the agreement, the company's existing standard product could measure up to four channels of torque/tension (strain gauge-based), the angle of bolt/fastener rotation, and other analog input channels, but lacked the ultrasonic element and the ability to perform all of these functions at the same time. By licensing NASA technology, Micro Control integrated the ultrasonic measurement aspect into its standard product, which enabled it to measure bolt tension directly using standard fasteners, and acquired the knowledge to measure torque/tension and angle rotation simultaneously.

The new MC900 Transient Recorder Analyzer provides fastener engineers with a powerful tool for studying threaded fastener joint designs or dynamic analysis of nut-runner operations on the plant floor or in a laboratory environment. The biggest challenge for a fastener engineer working in these areas is determining the clamp force between the two parts that are intended to be fastened together. In the past, the only information available to the engineer was the bolt's torque and angle of rotation. This required the engineer to establish a relationship between torque and tension. Based on this relationship, a determination could be made on the torque strategy that would be used in production.

The problem with this method, however, is that slight changes in thread friction or bolt geometry from one bolt to another can cause substantial errors between applied torque or rotation and the actual clamping force of the fastener. With the MC900 analyzer, data analysis and collection becomes simplified for the engineer, thereby diminishing the challenge of the clamp force process. Engineers can use the tool to ultrasonically measure the clamping force of the bolted joint during production without the need for specially fabricated bolts instrumented with strain gauges.

According to Micro Control, conventional ultrasonic testing methods can only measure elongation/load, based on the theoretical values and geometry of a bolt. Though in reality, the materials in the bolt rarely match exactly to the published standard materials, and the geometrical complexities of the bolt make it difficult to decipher accurate elongation. Additionally, the theoretical method only works in the elastic region of the bolt.

In contrast, MC900 allows for analyzation of the theoretical values and geometry of the bolt; calibration of a sample of bolts with a known load, using a load cell or tensile machine and then using the averaged results; or calibration of each bolt individually by pulling the bolts with the tensile machine. An accurate relationship between elongation and load is established, and the user can witness the performance of the bolt beyond the elastic region, and into the plastic region.

For determining ultrasonic delays and strain in a bolt, MC900 employs Langley's "pulsed phase-lock loop" technology. This system sends a toneburst through an ultrasonic transducer, which transmits a series of high-frequency sound waves into a specimen (the NASA technology utilizes just one frequency of sound waves, as opposed to the broad frequency spectrum occupied by traditional ultrasonic tension measurements). The sound waves' echo is then received from the far end of the bolt, and the phase shift is computed by comparing the phase of the returned signal with that of the original toneburst as the bolt is tightened. To identify the specific cycle in the return echo for elongation measurement, MC900 applies pattern matching, keeping the reference echo in memory for subjective comparison with the final echo. Each of the device's four channels can be programmed as "pitch catch"—with transducers on both ends of the bolt, "pulse echo"—with a single transducer on one end of the bolt, or "multiple echo," depending on the physical properties of the bolts to be measured.

MC900's powerful hardware/Microsoft® Windows®-based software combination comes fully loaded with user options, including additional tuning filters to accommodate different ultrasonic transducers, ID recognition for smart sensors, statistical and graphical analysis in real time, cross-plotting, automatic or manual calibration, and unlimited recording time of data.

Micro Control also developed and patented a low-cost, reusable "glue on" ultrasonic tension sensor called a UTensor™ that can be used in conjunction with the MC900 device to produce the accuracy and repeatability heavily needed in the automotive industry. By gluing the UTensor onto a fastener, the user can measure the preload, or relaxation point, after tightening a bolt.

Currently, "big three" automakers General Motors Corporation, Ford Motor Company, and DaimlerChrysler, along with their suppliers and bolt manufacturers, are incorporating the MC900 and UTensor technology.

Micro Control, Inc.'s UTensor,™ a low-cost, reusable "glue on" ultrasonic tension sensor, provides dynamic reading of tension while a bolt is being tightened.

Creating With Carbon

By combining the first variety of X-ray tubes ever available with a carbon-based nanotube innovation born just over a decade ago, Applied Nanotech, Inc., is helping to put the nanotechnology field on the map with promising new advances aimed at revolutionizing medicine, television, and everything in between.

A subsidiary of SI Diamond Technology, Inc., Applied Nanotech, of Austin, Texas, is creating a buzz among various technology firms and venture capital groups interested in the company's progressive research on carbon-related field emission devices, including carbon nanotubes, filaments of pure carbon less than one ten-thousandth the width of human hair. Since their discovery in 1991, carbon nanotubes have gained considerable attention due to their unique physical properties. For example, a single perfect carbon nanotube can range from 10 to 100 times stronger than steel, per unit weight. Recent studies also indicate that the nanotubes may be the best heat-conducting material in existence. These properties, combined with the ease of growing thin films or nanotubes by a variety of deposition techniques, make the carbon-based material one of the most desirable for cold field emission cathodes.

Image courtesy of Oxford Instruments plc.

The Horizon600 hand-held spectrometer can perform analyses on 20 different chemical elements within seconds.

Even more, the carbon nanotube may just one day be the choice composite used to bear the weight of NASA's fantastic future space travel concept, the space elevator, which would serve as a low-energy, mass transportation system for space-bound personnel, satellites, and other payloads. Although development of a space elevator could take as many as 100 years to successfully accomplish, NASA scientists are currently mulling over the possibilities of using carbon nanotubes to produce the elevator's long cable. This cable would be attached to a point on the Earth's equator, and extend into space, with its center of mass at geostationary Earth orbit.

Applied Nanotech is blending this next-generation technology of unparalleled strength and conductivity with cold cathode X-ray tubes, which were first introduced in the late 19th century, only to vanish quickly with the invention of a hot cathode called ductile tungsten in 1910. With support from NASA's Jet Propulsion Laboratory (JPL) under a **Small Business Innovation Research (SBIR)** contract, Applied Nanotech was able to bring cold cathode technology back to life with the mindset to produce smaller and more energy-efficient X-ray devices, especially ideal for applications requiring minimum power within harsh environments. NASA is considering replacing the standard hot cathodes used on satellites with the carbon cold cathode technology as a source for low power electric propulsion thrusters. Applied Nanotech's collaboration with JPL has allowed the company to extend the applications of its cold cathode technology and enhance its licensing policies.

For its first commercial achievement involving carbon cold cathodes, Applied Nanotech customized and sold the technology to Oxford Instruments plc—based in the United Kingdom, with American offices in Concord, Massachusetts, and Clearwater, Florida—for use in the company's Horizon600 self-contained, portable X-ray fluorescence spectrometer. The Horizon600's unique digital cold cathode X-ray tube allows for pinpoint therapy in medical fields such as brachytherapy, in which intra-coronary radiation is used to relieve blockages and prevent restenosis in stents, and dentistry, all at a lower cost and greater safety than conventional methods, due to the absence of radioactive isotopes. With its ability to perform analyses on a range of 20 different chemical elements within seconds, the hand-held device is also designed to detect asbestos, lead, and other dangerous contaminants hidden within walls, floors, and ceilings.

The ergonomically designed Horizon600 instrument features a color touch screen display for ease of use, a multi-channel digital signal processor for quick, stable results, and a built-in Microsoft® Windows® Pocket PC operating system complete with analytical software and an SQL database for results storage. It runs on a rechargeable, high power battery, allowing for extensive operation time, short "warm-up" for instant measurements, and overall low power consumption. In addition to Oxford Instruments, Applied Nanotech is supplying carbon nanotube cold cathodes to MediRad of Ra'anana, Israel, and a Japanese company, both of which are leaders in developing new miniature X-ray tubes for medical applications.

Through its SI Diamond parent, Applied Nanotech also completed various SBIR projects with NASA's Glenn Research Center and Johnson Space Center that have influenced the company's understanding of high-definition picture element tubes (PETs). Applied Nanotech is now incorporating PETs, in conjunction with thin carbon films, in large, flat panel nanotube displays for indoor and outdoor lighted advertising billboards. The company believes that these types of displays will eventually take control of a market that is presently pushing plasma and liquid crystal displays as top-of-the-line technologies.

The carbon-based lighted displays developed under SI Diamond will be applied to the forthcoming VERSAtile™ electronic billboard product line, which is similar to the light-emitting diode (LED) displays seen in New York's Times Square, but at a quarter of the cost, according to the company. Unlike LED-based displays, however, SI Diamond's are completely viewable in direct sunlight, even from a 45-degree angle.

Applied Nanotech recently licensed its PET technology to two Japanese corporations, one of which is imaging solutions specialist Canon, Inc., in an effort to commercialize a new breed of superior, high-resolution, large-format televisions, otherwise known as "slim wall televisions." Manufacturing is expected to take place by 2004, with hopes that the technology will draw appeal from the American market.

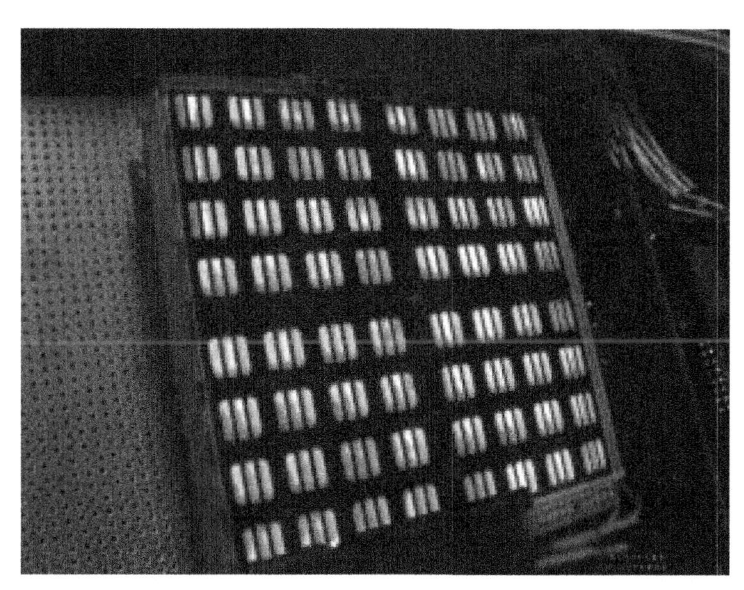

A picture element tube possessing NASA-influenced carbon cold cathode technology developed by Applied Nanotech, Inc. This full-color field emission display tube is 3 by 3 inches and approximately 1 centimeter in depth, and can be combined with many similar tubes to make billboard-size displays.

Thermographic Inspections Save Skins and Prevent Blackouts

To help ensure the safety of the Nation's aging fleet of commercial aircraft, NASA's Langley Research Center developed "scanning thermography" technology that nondestructively inspects older aircraft fuselage components. Scanning thermography involves heating a component's surface and subsequently measuring the surface temperature, using an infrared camera to identify structural defects such as corrosion and disbonding. It is a completely noninvasive and noncontacting process. Scans can detect defects in conventional metals and plastics, as well as in bonded aluminum composites, plastic- and resin-based composites, and laminated structures. The apparatus used for scanning is highly portable and can cover the surface of a test material up to six times faster than conventional thermography.

NASA scientists affirm that the technology is an invaluable asset to the airlines, detecting potential defects that can cause structural failure, such as that of Aloha Airlines Flight 243 in 1988. According to the National Transportation Safety Board, the Aloha Airlines accident was caused by the structural separation of the pressurized fuselage skin. As a result of this separation, "an explosive decompression occurred, and a large portion of the airplane cabin structure…was lost."

An extension of NASA's scanning thermography now offers considerable value to the Nation's utility companies, as nondestructive inspection methods are becoming an increasingly attractive means of determining the condition of critical components. This would include power and process plant machinery, roads and bridges, and building structures.

In 1996, ThermTech Services, Inc., of Stuart, Florida, approached NASA in an effort to evaluate the technology for application in the power and process industries, where corrosion is of serious concern. ThermTech Services proceeded to develop the application for inspecting boiler waterwall tubing at fossil-fueled electric-generating stations. In 1999, ThermTech purchased the rights to NASA's patented technology and developed the specialized equipment required to apply the inspecting method to power plant components.

The ThermTech robotic system using NASA technology has proved to be extremely successful and cost effective in performing detailed inspections of large structures such as boiler waterwalls and aboveground chemical storage tanks. It is capable of inspecting a waterwall, tank-wall, or other large surfaces at a rate of approximately 10 square feet per minute or faster. The inspection results provide a computerized map of the wall thickness with high-resolution data equivalent to that of existing inspection methods. Prior to the development of this technology, existing inspection methods would only allow inspection of a limited area (less than 10 percent) of the entire structure, because they are typically either manually operated or mounted on small-scale robots. Now, with the development of the thermal line scanner, it is possible to inspect nearly 100 percent of the structure in approximately the same time it takes to perform the limited area inspection.

ThermTech Services' system benefits the electric utility industry, saving utility customers millions of dollars by reducing maintenance costs and downtime and improving power plant reliability.

ThermTech Services, Inc.'s robotic crawler scales the inside wall of a boiler at the Schuylkill Generating Station, Philadelphia, Pennsylvania. During the inspection, the initial call outs indicated 54 tubes in the superheat furnace rear wall should be removed. Upon removal, the 54 tubes were subjected to a boroscope exam, and the defects found by the crawler were confirmed visually.

Precise Measurement for Manufacturing

A metrology instrument known as PhaseCam™ supports a wide range of applications, from testing large optics to controlling factory production processes. This dynamic interferometer system enables precise measurement of three-dimensional surfaces in the manufacturing industry, delivering speed and high-resolution accuracy in even the most challenging environments.

PhaseCam originated from a prototype interferometer that was being developed by MetroLaser, Inc., in 1999. During that time, Philip Stahl, a NASA engineer at Marshall Space Flight Center, learned of the technology while touring the company's facility, and immediately recognized its applicability to testing large astronomical mirrors and space optical systems. MetroLaser proposed building a system to NASA specifications for testing large optics in a vibrating environment. The technology would, among others, benefit NASA's Advanced Mirror System Demonstrator project for the James Webb Space Telescope.

In January 2000, 4D Vision Technology, Inc., was formed to commercialize the PhaseCam technology, with NASA becoming the firm's first customer. Just 6 months after NASA granted 4D Vision a contract, the company delivered its first PhaseCam system. Stahl stated that the company "took a task that was thought to be impossible and successfully accomplished it in less time and for less money than any of its competitors." As a result of the company's excellent work, NASA invited 4D Vision to present its new product at Technology Days 2001, an annual symposium held at Marshall to discuss the progress of various optics projects by NASA, contractors, and universities. This provided the company the opportunity to introduce PhaseCam to many potential customers in the commercial marketplace. In 2002, 4D Technology Corporation, of Tucson, Arizona, acquired 4D Vision as part of its mission to become a world leader in dynamic optical metrology products and services.

PhaseCam satisfies industry demands to produce accurate measurements where vibration and motion are intrinsic components of the manufacturing process, and yield and throughput are paramount. With this product, vibrations, moving parts, air turbulence, and other impediments are no longer a serious barrier to interferometric testing. Unlike phase-shifting interferometers, the system works by capturing data in a single frame, measuring data rates in tens of microseconds. By using a single camera to record four data frames at the exact same time, PhaseCam eliminates critical alignment issues and simplifies calibration. No matter how much vibration is present, all of the data represent the same instant in time.

Compact and reliable, PhaseCam enables users to make interferometric measurements right on the factory floor. The system can be configured for many different applications, including mirror phasing, vacuum/cryogenic testing, motion/modal analysis, and flow visualization. Customers include leading aerospace and optical manufacturers such as Eastman Kodak Company, Ball Aerospace & Technologies Corporation, and the University of Arizona Mirror Laboratory. NASA continues to use the technology to test mirror technologies for next-generation space telescopes. According to Stahl, "Not only did NASA get a great interferometer to enable the testing of large mirrors, but the taxpayer received great value. I believe that this type of proactive investment is an example of the government at its best."

PhaseCam™ is a trademark of 4D Technology Corporation.

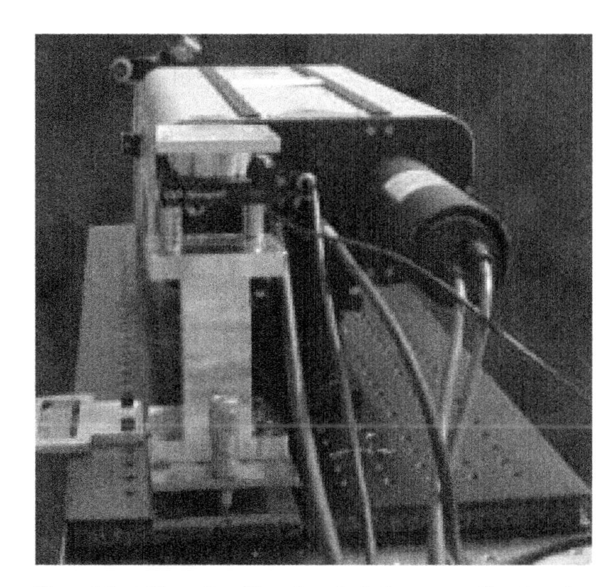

The original PhaseCam™ system tested a composite mirror for Marshall Space Flight Center.

Processing at the Speed of Light

Spatial Light Modulators (SLMs) are critical elements in optical processing systems used for imaging, displaying, data storage, communications, and other applications. By taking advantage of the natural properties of light beams, the devices process information at speeds unattainable by human operators and most machines, with high-resolution results.

Boulder Nonlinear Systems, Inc., is one of the world's foremost SLM manufacturers. Founded as a research and development firm by a group of University of Colorado researchers in 1988, Lafayette, Colorado-based Boulder Nonlinear essentially grew out of a **Small Business Innovation Research (SBIR)** contract with the U.S. Department of Defense/Air Force a year earlier. Founders Steve Serati, the company's current chairman and chief technical officer, and Gary Sharp submitted an SBIR proposal to improve the sensitivity of coherent light lidar systems using an optical technique to suppress spatial noise. Since winning that contract over 15 years ago, Boulder Nonlinear has benefited from more than $10 million in SBIR funding, with many contracts coming from NASA's Ames Research Center, Glenn Research Center, Jet Propulsion Laboratory (JPL), and Johnson Space Center.

With vast government support through the SBIR program and a strong knowledge of optics and liquid crystals, Boulder Nonlinear was able to establish itself as a successful custom-manufacturing entity with a reputable SLM product. Its latest adaptation of government-influenced technology is the 512x512 Multi-Level/Analog Liquid Crystal SLM. The high frame rate device, developed under SBIR contracts with Ames, Glenn, and JPL, modulates light in pure amplitude, pure phase, or coupled amplitude and phase. To achieve the desired optical response, the 512x512 SLM can be filled with either a Ferroelectric Liquid Crystal, best used for applications requiring very fast modulation along the real-axis (i.e., amplitude only) or for pure phase modulation of $\pi/2$ or less, or a Nematic Liquid Crystal, best used for pure phase modulation of 2π or for applications not requiring pure real-axis modulation. All 512x512 devices are custom designed to yield the maximum modulation response at the desired wavelength.

Boulder Nonlinear also offers its customers an entry-level 256x256 SLM that delivers high-speed frame rates and sharp resolution for smaller, cost-effective

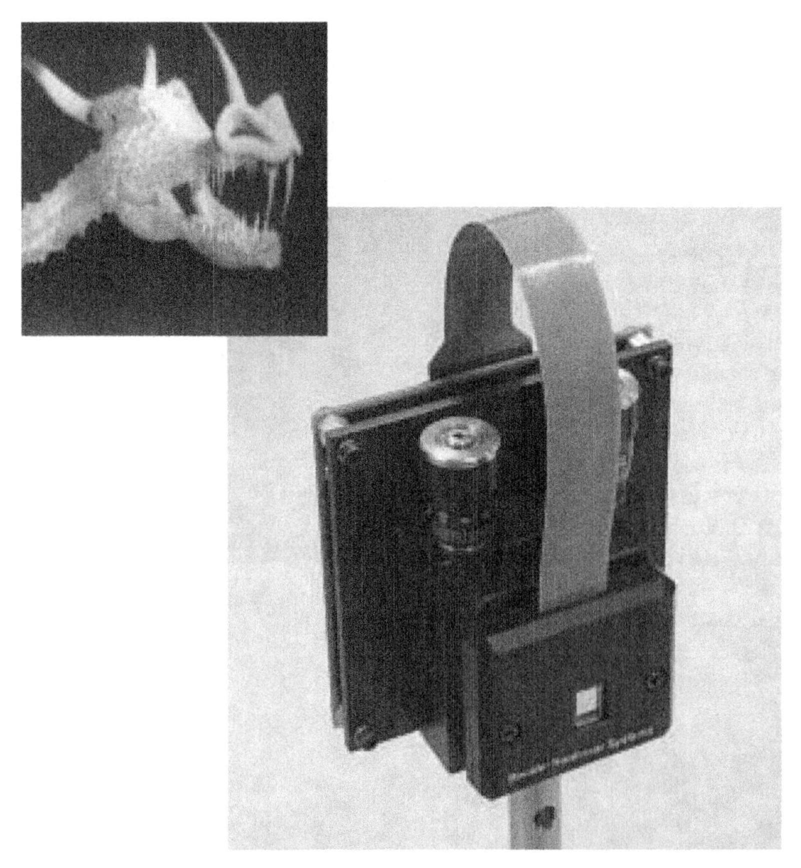

Boulder Nonlinear Systems' liquid crystal Spatial Light Modulators, available in a variety of resolutions, are ideal for optical correlation, beam steering, telecommunications, defense, and medical research applications. The grayscale image of the dragon was taken on a Ferroelectric Liquid Crystal 512x512 Spacial Light Modulator.

Without touching: A molecule is lifted by "light tweezers." This unique technology, developed by Boulder Nonlinear Systems in partnership with a client, is noninvasive and eliminates the risk of contamination.

applications. With help from Ames, the company is currently working on commercializing a 1024x1024 SLM, the next addition to its highly regarded SLM product line. The company also markets a 1x4096 SLM that received the prestigious Photonics Circle of Excellence Award as one of the 25 most innovative photonics products introduced in 2000. The device is primarily used as an optical phased array (OPA) for nonmechanically steering a laser beam. Through a Phase II SBIR award, Langley Research Center is currently funding a next-generation OPA that will increase scanning speed, scanning range, and the active area.

Other applications for Boulder Nonlinear's SLM line include medical research, forensics, laser printing and scanning, holography, and laser beam steering (refractive and diffractive). In medical research, Boulder Nonlinear teamed up with a client to develop "light tweezers," a unique, noninvasive technology that can lift molecules or microscopic specimens without the risk of damage. In the forensics field, original equipment manufacturers have purchased the 512x512 SLM to build optical correlators to quickly analyze reams of visual evidence, such as fingerprints, to reveal known suspects. The 512x512 SLM is crucial for high-speed processing so that forensic experts can compare a known visual pattern to an unknown visual pattern in time-sensitive situations.

Look Dynamics, located in Longmont, Colorado, incorporated Boulder Nonlinear SLMs into its optical computer product that uses light, rather than electrons, to process information such as satellite images of Earth and images of the human body. Boulder Nonlinear states that the marriage of computer logic and liquid crystal beam manipulation "creates a new universe of possibilities for manipulating light to measure, analyze and inform." Other commercial and government entities have counted on Boulder Nonlinear SLMs and liquid crystal components to replace prisms, detect bombs and track missiles (target recognition), store massive data sets on credit cards, link mobile stations to a broadband network, recognize faces, and see objects in turbulent conditions.

In its latest SBIR venture with NASA, Boulder Nonlinear is helping JPL develop SLM technology for small body exploration and Mars sample return missions. The project may also lead to new means of location recognition for spacecraft docking and landing.

Keeping Communication Continuous

General Dynamics Decision Systems employees have played a role in supplying telemetry, tracking, and control (TT&C) and other communications systems to NASA and the U.S. Department of Defense for over 40 years. Providing integrated communication systems and subsystems for nearly all manned and unmanned U.S. space flights, the heritage of this Scottsdale, Arizona-based company includes S-band transceivers that enabled millions of Americans to see Neil Armstrong and hear his prophetic words from the Moon in 1969. More recently, Decision Systems has collaborated with NASA's Goddard Space Flight Center to develop transponders, wireless communications devices that pick up and automatically respond to an incoming signal, for NASA's Tracking and Data Relay Satellite System (TDRSS).

Four generations of Decision Systems' TDRSS transponders have been developed under Goddard's sponsorship. The company's Fourth Generation TDRSS User Transponder (TDRSS IV) allows low-Earth-orbiting spacecraft to communicate continuously with a single ground station at White Sands, New Mexico, through a constellation of geostationary relay satellites positioned at key locations around the Earth. In addition to the communications of forward link control commands and return link telemetry data, the TDRSS IV also supports spacecraft orbit tracking through coherent turn-around of a pseudo-noise ranging code and two-way Doppler tracking.

Up until now, the highly successful relationship between Goddard and Decision Systems had produced a high quality product used primarily for NASA-specific programs. This changed when the National Scientific Balloon Facility (NSBF), in Palestine, Texas, turned to Decision Systems in need of a TT&C communications system for its high-altitude, long-duration balloon mission program.

Because many of the NSBF's balloon missions are flown at the Earth's poles, where commercial communications services are not readily available, the continuous communications feature provided by NASA's TDRSS is extremely important during a long mission. When the NSBF adopted the use of global positioning system receivers for balloon position tracking, Decision Systems concluded that a simpler, noncoherent transceiver could provide the NSBF with the necessary TDRSS communications without the additional cost and complexity of a coherent transponder. The solution was to take the core design of the TDRSS IV Transponder, but remove the extra functionality that supported coherent turn-around. This would simplify the production effort, reduce the testing required, and result in a lower cost product with smaller size, weight, and power consumption.

Once NSBF and Decision Systems agreed on a concept for this new product, known as the Multi-Mode Transceiver (MMT), the NSBF approached Goddard for approval and funding. The expertise and cooperation of Goddard engineers was critical during the test and evaluation phase of the device's production process, and the first MMT underwent TDRSS compatibility testing at Goddard before any of the commercial production units could be delivered.

With the new MMT's reduced size, weight, power consumption, and cost, the advantages of TDRSS communications are now available for a variety of applications outside of the traditional NASA spacecraft missions. With the MMT, the NASA system is now accessible to university satellite programs, small commercial Earth imaging programs, as well as Arctic and Antarctic science programs.

The Multi-Mode Transceiver brings the advantages of NASA's Tracking and Data Relay Satellite System to a variety of applications, including university satellite programs, small commercial Earth imaging programs, and Arctic and Antarctic science programs.

Designing It Smart With SIV

When research staff at NASA's Glenn Research Center developed and patented Stereo Imaging Velocimetry (SIV), the world's first three-dimensional (3-D), full-field quantitative and qualitative analysis tool to investigate flow velocities, experiments that were previously impossible became a reality. Seizing the opportunity to commercialize NASA's breakthrough invention, Digital Interface Systems (DIS), Inc., of North Olmsted, Ohio, acquired an exclusive license to market SIV, which has a range of applications from improving the aerodynamics of aircraft and automobiles to avoiding "no flow" regions in artificial hearts.

NASA Glenn's SIV is a safe, affordable means to obtain 3-D flow information from any transparent liquid that can be seeded with tracer particles. Previously, accurate information of this type was very difficult to obtain and often required using dedicated laser-based measurement systems. Eliminating the need for lasers, SIV provides an effectively nonintrusive measurement of 3-D fluid velocities at many points and at high frame speeds using two charge coupled device (CCD) video cameras and neural networked-based computational algorithms. It allows for the direct comparison of computed and experimentally measured fluid flows, with no limitation on the fluid flow scale to be measured.

The five distinct steps in the SIV method include camera calibration, centroid/overlap decomposition, particle tracking, stereo matching, and 3-D analysis. The CCD cameras are oriented at 90° with respect to each other in order to observe a fluid experiment that has been seeded with the small tracer particles. Each camera records two-dimensional (2-D) data of the particles' motion in the observation volume. Users obtain the 3-D data by computationally combining the 2-D information from both cameras. The SIV method incorporates a camera-calibration technique in which rotation and translation of camera lenses and optical distortion generated in the lenses are taken into account using the accurate 2-D- to 3-D-mapping function.

SIV applies to diverse experiments such as the study of multiphase flow, bubble nucleation and migration, pool combustion, and crystal growth. The technique successfully analyzed data from two Space Shuttle missions. Several of NASA's ground-based experiments are also benefiting, as SIV is applied to the Agency's microgravity program for fluid physics experiments.

In the commercial marketplace, SIV applies to industrial process optimization and the design of new products. It helps companies to create more efficient heating, ventilating, and air conditioning systems, as well as quieter airflow within auto heating and cooling ducts. The technology can assist with air flow studies around buildings and the modeling of continuous casting operations. SIV has been used in the steel industry to quantify the continuous casting process, the vacuum cleaner industry to observe brushroll designs, and in the sporting goods industry to investigate the bat-ball impact phenomenon of softball bats.

SIV is available through DIS as an on-demand World Wide Web deployable program or as a mini-compact disc version, with robust, user-friendly, graphical user interface enhancements that enable easy navigation of the tool.

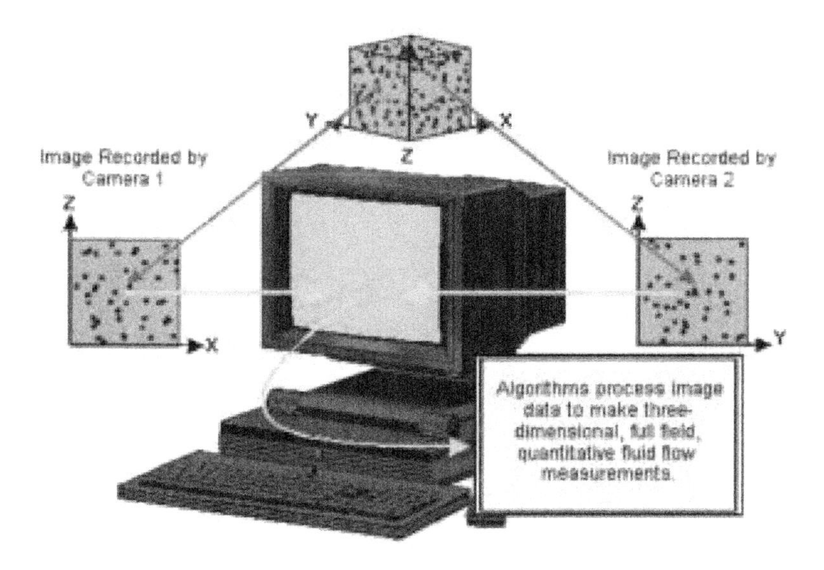

This diagram shows a fluid experiment seeded with tracer particles. Two cameras are set up perpendicular to each other, recording two widely disparate views. The two views are computationally combined to obtain three-dimensional coordinates of the seed particles.

NASA
802

One Hundred Years of Powered Flight

NASA is proud of its achievements, as well as those of its predecessor, the National Advisory Committee for Aeronautics, during the first century of powered flight. This year's "Centennial of Flight" celebration offers a unique opportunity for NASA's 10 field centers to showcase the Agency's historical and ongoing contributions to aeronautics. Through advanced scientific research and technology transfer, NASA continues to impact flight through improved safety, efficiency, and cost effectiveness.

One Hundred Years of Powered Flight

This year, Centennial of Flight celebrations across the United States are marking the tremendous achievement of the Wright brothers' successful, powered, heavier-than-air flight on December 17, 1903. The vision and persistence of these two men pioneered the way for explorers, inventors, and innovators to take aeronautics from the beaches of Kitty Hawk, North Carolina, to the outer reaches of the solar system. Along this 100-year journey, NASA has played a significant role in developing and supporting the technologies that have shaped the aviation industry.

In this photo taken on March 15, 1929, NACA staff members conduct tests on airfoils in the Variable Density Tunnel. In 1985, the Variable Density Tunnel was declared a National Historic Landmark.

NASA's Origin

NASA's commitment to aeronautics technology stems back to 1915 when Congress established the National Advisory Committee for Aeronautics (NACA) through a rider to the Naval Appropriations Act. Much of what the United States takes for granted in aviation today was pioneered by NACA, which was transformed into NASA under the National Aeronautics and Space Act of 1958. NACA made advancements and contributions to every field associated with aeronautics, including the fledging field of spaceflight.

In the beginning, NACA concentrated on problems related to military aviation, spurred by the onset of World War I. NACA scientists and engineers, charged with the mission "to supervise and direct the scientific study of the problems of flight, with a view to their practical solution," developed a strong technical competence and a commitment to collegial in-house research conducive to engineering innovation. When the war ended, NACA engineers turned their attention to solving a broad range of problems in flight technology. At the first NACA Aircraft Engineering Research Conference in 1925, the U.S. Government promoted the transfer of technology and expertise to industry.

The official seal for National Advisory Committee for Aeronautics depicts the first human-controlled, powered flight made by the Wright brothers in December 1903 at Kitty Hawk, North Carolina.

Establishing a Tradition of Excellence

Over the years, NACA expanded to conduct its research at three major research laboratories—the Langley Memorial Aeronautical Laboratory established in 1917, the Ames Aeronautical Laboratory formed in 1940, and the Lewis Flight Propulsion Laboratory founded in 1941. NACA also maintained a small Washington, DC, headquarters and two small test facilities.

One of the earliest landmark events for NACA was the formal dedication of the wind tunnel at NACA Langley, later renamed NASA Langley Research Center. This tunnel was the first of many now-famous NACA/NASA wind tunnels, enabling engineers and scientists to develop advanced wind tunnel concepts to support aircraft design. In 1922, Langley's Variable Density Wind Tunnel (VDT) employed high-pressure air to better simulate flight conditions with scale models. The VDT was the most advanced wind tunnel of its day, helping the United States become a leader in aeronautical research. Over the next 10 years, NACA compiled research from the VDT that culminated into NACA Report 160, which included a groundbreaking systematic study of airfoils and produced the airfoil numbering system of today.

Langley metal workers fabricated NACA cowlings for early test installations.

NACA continued to expand by starting the Cowling Wind Tunnel research at Langley in 1928. The research paved the way for a series of component drag studies, helping NACA to develop a low-drag cowling for radial air-cooled aircraft engines. This breakthrough technology was adopted by all aircraft manufacturers, as the cowling greatly reduced the drag that an exposed engine generated, resulting in significant cost savings and increased aircraft speeds and range. The National Aeronautic Association awarded NACA the Robert J. Collier Trophy, the most prestigious award for great achievement in aeronautics and astronautics in America, for the low-drag cowling. This would be the first of 19 Collier Trophies for NACA/NASA leading up to the present day.

Strategic Value in World War II

When NACA Langley's Atmospheric Wind Tunnel became operational in 1930, it produced a knowledge base and essential design data relative not only to basic aircraft performance, but also to aircraft stability and control, power effects, flying qualities, aerodynamic loads, and high-lift systems. While NACA's research had both military and civil applications prior to the outbreak of World War II, its activity became almost exclusively military with stronger industry ties during the war. Dozens of corporate representatives visited Langley during this time to observe and assist in testing, as NACA focused on refining and solving specific problems. One major advance was the development of the laminar-flow airfoil to solve a turbulence problem at the wing trailing edge that was limiting aircraft performance.

Langley also helped to improve the performance of existing aircraft through tests in its full-scale tunnel and 8-foot, high-speed tunnel. The 8-foot tunnel was unlike any other in the world, giving the United States a strategic advantage in the war. The first tests in the tunnel evaluated the effects of machine gun and cannon fire on the lift and drag properties of wing panels. This led engineers to check the effects of rivet heads, lapped joints, slots, and other irregularities on drag. The tests demonstrated drag penalties as high as 40 percent over aerodynamically smooth wings. As a result, aircraft manufacturers quickly switched to flush rivets and joints.

New high-speed propellers and engine cowlings also emerged from tests in Langley's 8-foot tunnel, but the development of the Lockheed P-38 Lightning dive recovery flap provided tremendous proof of the wind

The 8-foot, high-speed wind tunnel at the NACA Langley Aeronautical Laboratory provided the means for testing large models and some full scale components at a simulated speed of 500 miles per hour.

View of the 16-foot, high-speed wind tunnel at the NACA Ames Aeronautical Laboratory in Moffett Field, California.

tunnel's value. The P-38, a high-speed, twin-boom fighter that helped beat back the threat of Japanese Zero airplanes in the South Pacific, introduced a new dimension to American fighters with its second engine. The multi-engine configuration reduced the P-38 loss-rate to anti-aircraft gunfire during ground attack missions.

When the P-38 was first introduced into squadron service in 1941, pilots were plagued by heavy buffeting during high-speed dives. On several occasions, their dives steepened and they could not pull out. Lockheed's test pilot for the P-38, Ralph Virden, lost his life trying to solve the dive problem. Shortly after Virden's death, the Army asked NACA for help. Crucial tests were conducted using one-sixth scale models in the 8-foot tunnel, indicating that above 475 miles per hour, the P-38's wings lost lift and the tail buffeted, causing a strong, downward pitching motion of the plane. Controls stiffened up, preventing the pilot from pulling the plane out of its dive. In addition, the buffeting could cause structural failure, as it had in Virden's case.

Langley's solution to the P-38 dive problem was the addition of a wedge-shaped dive recovery flap on the lower surface of the wings. Aerodynamic refinement of the dive recovery flap was continued in a coordinated program between Lockheed engineers and NACA's newly founded Ames Aeronautical Laboratory (later renamed NASA Ames Research Center) in Moffett Field, California, in the latter's 16-foot, high-speed tunnel. The dive recovery flaps ultimately were incorporated on the P-47 Thunderbolt, the A-26 Invader, and the P-59 Airacobra, America's first jet aircraft.

Aviation Safety Measures

Right from the start, the aviation community was unanimous in its desire to develop systems and procedures that would make flying safer. While NACA information on airflow over wing surfaces and dive flaps helped pilots retain control over diving airplanes, NACA was also asked to determine how air crews and aircraft could better withstand water impacts. NACA forwarded test results regarding the problem to aircraft manufacturers, resulting in new designs that helped save the lives of countless air crews.

Another concern for aviation safety was ice build-up on aircraft wings and propellers, which reduced lift and increased drag, leading to fatal crashes. From 1936 into the mid-1940s, NACA created Thermal Ice Prevention Systems to investigate effective countermeasures to the problem of ice formation on aircraft. Ames, in particular, began to make strides in icing research. Prototypes of an Ames anti-icing system were evaluated in the B-17 and B-24 heavy bombers during World War II, while a substantive program on the C-46A Commando icing research aircraft led to the definition of icing system design criteria.

Ames' icing work consisted of both research and extensive design of actual hardware installed on airplanes. In 1946, Lewis A. Rodert, a leading icing researcher at Ames, received the Collier Trophy for the development of an efficient wing deicing system, which piped air heated by hot engine exhaust along the leading

The Curtiss C-46A Commando icing research aircraft helped Ames define icing system design criteria.

This mid-1950s photograph shows the Douglas D-558-2 and the North American F-86 Sabre chase aircraft in flight. Both aircraft display early examples of swept wing airfoils.

edge of the wing. The system protected the lives of many pilots flying in dangerous weather conditions.

In addition to icing research, Ames worked on transonic variable stability aircraft and flying qualities, stability and control, and performance evaluations. Flight research at Ames progressed from idea development to stages of wind tunnel and ground-based simulator tests to analyses in its facilities. Collaborative efforts to use a combination of facilities led to more substantive results. The standardized system for rating an aircraft's flying qualities is perhaps the most important contribution of the evaluation programs and experiments conducted on the variable stability aircraft at Ames. George Cooper, Ames' Chief Test Pilot, developed the rating system to quantify the pilot's judgment of an aircraft's handling and apply it to the stability and control design process.

Cooper's approach forced a specific definition of the pilot's task and performance standards. Further, it accounted for the demands the aircraft placed on the pilot in accomplishing a given task to some specified degree of precision. The Cooper Pilot Opinion Rating Scale was initially published in 1957. After gaining several years of experience through applying the scale to many flight and simulator experiments and through its use by the military services and aircraft industry, Cooper modified it in collaboration with Robert Harper of the Cornell Aeronautical Laboratory in 1969. The Cooper-Harper Handling Qualities Rating Scale is one of the enduring contributions of Ames' flying qualities research, as the scale remains the standard for measuring flying qualities to this day.

Picking up Speed

In 1944, Ames' 40- by 80-foot full-scale wind tunnel became operational, allowing whole aircraft to be wind tunnel tested, as compared to models at low-flight speeds. As World War II ended in 1945, NACA aerodynamicist Robert T. Jones developed the Swept Wing Concept, which identified the importance of swept-back wings in efficiently achieving and maintaining high-speed flight. Jones verified the concept, designed to overcome shockwave effects at critical Mach numbers, in wind-tunnel experiments. To this day, Jones' swept-back wings are used on almost all commercial jet airliners and military craft.

High-speed flight research was often a collaboration between NACA and U.S. Army Air Forces. In the late 1940s and throughout the 1950s, a succession of experimental aircraft was flown at Muroc Army Airfield (now Edward's Air Force Base) through NACA's Muroc Flight Test Unit, which would later become NASA's Dryden Flight Research Center. On November 14, 1947, the air-launched, rocket-powered X-1 aircraft broke the sound barrier. Chuck Yeager piloted the X-1 during this monumental flight, ushering in the era of supersonic flight. Record flights by the military and NACA's rocket planes probed the characteristics of high-speed aerodynamics and stresses on aircraft structures. NACA's John Stack led the development of a supersonic wind tunnel,

The X-1 was the first piloted supersonic aircraft to break the sound barrier.

Glenn Research Center's Icing Research Tunnel was established in 1944. The Altitude Tunnel is in the center background, the Propeller Motor Drive Housing is in the right background, and the Air Dryer and Cooling Tower is in the left background.

speeding the advent of operational supersonic aircraft. He shared the Collier Trophy in 1947 with Yeager and Lawrence Bell for research determining the physical laws affecting supersonic flight.

Meanwhile, NACA's newly founded Aircraft Engine Research Laboratory in Cleveland, Ohio—today's NASA Glenn Research Center—was steadily translating German documents on jet propulsion tests that quickly became basic references in the new field of gas turbine research. Italian and German professionals joined their American colleagues at Glenn to work on this new aspect of flight research. To cope with continuing problems of how to cool turbine blades in the new turbojets, Glenn's Ernst Eckert laid the basic foundation for heat transfer research as the laboratory examined this issue.

In 1944, Glenn also began testing ice protection systems with the completion of its Icing Research Tunnel. Most ice protection technologies in use today were largely developed at this facility. In 1987, the American Society of Mechanical Engineers designated Glenn's tunnel an International Historic Mechanical Engineering Landmark for its leading role in making aviation safer for everyone.

Between 1950 and 1960, Glenn engineers pursued the development of an axial flow compressor for jet engines that improved efficiency by an order of magnitude. This research became the basis of the modern high-bypass jet turbofan. The engineers also developed new ways to solve complex combustion, heat exchange, and supercharger problems. While engine research did not receive much public attention at this time, Glenn's work on Army aircraft contributed to one of the military's most successful airplanes—the Army's Boeing B-17 Flying Fortress, a high-altitude, high-speed bomber.

In 1951, Langley's Richard T. Whitcomb determined the transonic "Area Rule," which explains the physical rationale for transonic flow over an aircraft. This concept allows modern supersonic aircraft to penetrate the sound barrier with greatly reduced power, and is now used in all transonic and supersonic aircraft designs. Twenty years later, Whitcomb went on to develop the supercritical wing, which yields improved cruise economy of approximately 18 percent by delaying the drag increase at transonic speeds, delaying buffet onset, and increasing lift. Whitcomb's concept is widely used on commercial and military aircraft today.

Another step in laying the groundwork for modern aviation was NACA Report R1135, "Equations, Tables, and Charts for Compressible Flow." Published in 1953, the report became a "bible" for compressible flow aerodynamics. Also that year, the D-558-2 aircraft, flown by Dryden's Scott Crossfield, was the first aircraft to break Mach 2, or twice the speed of sound. The achievement culminated a joint Navy/NACA high-speed flight research program.

The Dawn of NASA

NACA began researching flight beyond Earth's atmosphere in the early 1950s. Laboratories studied the possible problems of space flight, while engineers discussed aircraft that could reenter the atmosphere at a high rate of speed, producing a great amount of heat. Ames' H. Julian Allen developed the "blunt nose principle," suggesting that a blunt shape would absorb only a very small fraction of the heat generated by an object's reentry into the atmosphere. The principle was later significant to NASA's Mercury capsule development.

NACA's work also led to the creation of the rocket-propelled X-15 research airplanes, which helped verify theories and wind tunnel predictions to take piloted flight to the edge of space. Over the course of a decade, the X-15 program embarked on a new frontier, exploring the possibilities of a piloted, rocket-powered, air-launched aircraft capable of speeds around five times that of sound. The remarkable X-15—half plane, half rocket—bridged the gap between air and space flight, putting Dryden Flight Research Center on the map. Dryden's testing of the program's 199 flights produced invaluable data on aerodynamic heating, high-temperature materials, reaction controls, and space suits.

In 1957, when the Soviet Union launched Sputnik, the first man-made object to orbit the Earth, the United

In the 8-foot, high-speed wind tunnel in April 1955, Richard Whitcomb examines a model designed in accordance with his transonic Area Rule.

States' efforts to reach space intensified. On January 31, 1958, a U.S. Army rocket team at the Redstone Arsenal in Huntsville, Alabama, led by Army Ballistic Missile Agency Technical Director Dr. Wernher von Braun, launched a four-stage Jupiter-C rocket from a Florida launch site. The rocket carried Explorer I, the Nation's first Earth-orbiting satellite, and marked the United States' initial entry in the space race.

Following Explorer I, American leadership questioned whether the emerging U.S. Space Program should be administered by a military or civilian agency. The debate resulted in the creation of the National Aeronautics and Space Administration (NASA), a civilian organization, on October 1, 1958. When President Dwight D. Eisenhower signed the National Aeronautics and Space Act of 1958, all NACA activities and facilities were folded into the newly formed NASA. NACA and other organizations from the Army and Navy became the nucleus of the new agency, which was tasked with both aeronautics and astronautics responsibilities. While NASA's major focus would be space research, aeronautics remains the first "A" in its name.

Emerging NASA Centers Shoot for the Moon

President Eisenhower signed an executive order indicating that personnel from the Development Operations Division of the Army Ballistic Missile Agency in Huntsville should transfer to NASA. Soon after, in 1960, the Marshall Space Flight Center was founded to provide launch vehicles for NASA's exploration of outer space. Dr. von Braun became director of the new center in Huntsville, as he and his rocket team expanded the Jupiter C's payload lifting capabilities, creating the Juno II space vehicle to launch various Earth satellites and space probes.

During its first 5 years, NASA continued to expand its facilities. In 1959, Goddard Space Flight Center in Greenbelt, Maryland, was established as NASA's first

As crew members secure the X-15 rocket-powered aircraft after a research flight, the B-52 mothership used for launching this unique aircraft does a low fly-by overhead. Information gained from the highly successful X-15 program contributed to the development of the Mercury, Gemini, and Apollo piloted spaceflight programs, and also the Space Shuttle Program.

The swing arms move away and a plume of smoke signals the liftoff of the Apollo 11 Saturn V Space vehicle and astronauts Neil A. Armstrong, Michael Collins, and Edwin E. Aldrin, Jr. from Kennedy Space Center Launch Complex 39A.

space flight center. Under Goddard project management, the Explorer VI provided the world with its first image of Earth from space. In 1961, Houston, Texas, became the site of NASA's Manned Spacecraft Center, which was renamed Johnson Space Center in 1973 to honor the late President and Texas native, Lyndon B. Johnson.

NASA also announced its decision to build a national rocket test site in 1961, establishing the John C. Stennis Space Center in Hancock County, Mississippi, 3 years later to test the first and second stages of the powerful Saturn V rocket that Marshall developed for the Apollo and Skylab Programs. On July 1, 1962, the Marshall Launch Operations Directorate came to Florida to initiate NASA's Launch Operations Center, later renamed the John F. Kennedy Space Center. The Jet Propulsion Laboratory, managed by the California Institute of Technology, was also founded during the 1960s, leading NASA's robotic exploration of the universe.

On July 20, 1969, U.S. astronauts walked on the Moon for the first time after the Apollo 11 crew was launched from Kennedy and safely transported by Marshall-developed rocket boosters that were tested and proven flight worthy at Stennis. Mission Commander Neil Armstrong sent the message back to Johnson, "Houston, Tranquility Base here. The Eagle has landed!" With the completion of the Apollo Program in December 1972, NASA journeyed forward to build new aeronautic achievements upon its groundbreaking accomplishment.

Advancing Towards Modern Aviation

With the new frontier of space within its grasp, NASA continued to improve aviation for Earth-bound purposes. Initiated in 1972, NASA's Digital Fly-By-Wire (DFBW) flight research project, a joint effort between Dryden and Langley, validated the principal concepts of all-electronic flight control systems. On May 25, 1972, Dryden's highly-modified F-8C DFBW research aircraft, with pilot Gary Krier at the controls, became the world's first aircraft to fly completely dependent upon an electronic flight control system. Soon after, electronic fly-by-wire systems replaced older hydraulic control systems, freeing designers to create safer, more maneuverable, and more efficient aircraft.

NASA's DFBW system is the forerunner of current fly-by-wire systems used on the Space Shuttles and on military and civil aircraft such as F/A-18 fighters and the Boeing 777. Modern digital flight control systems make flying safer for both civil and military aircraft. Multiple computers "vote" instantaneously to choose the correct control input for maneuvers requested by the

The F-8C Digital Fly-By-Wire Control System was first tested in 1972. The use of electrical and mechanical systems to replace hydraulic systems for aircraft control surface actuation was flight-tested. Today widely used by commercial airliners, the system allows for better maneuver control, smoother rides, and for military aircraft, a higher combat survivability.

The Boeing Quiet Short-Haul Research Aircraft.

pilot, who uses the traditional stick and rudder controls in the cockpit. Digital systems make aircraft more maneuverable because computers command more frequent adjustments than human pilots. For airliners, computerized flight controls ensure a smoother ride than a pilot alone could provide with stick and rudder controls. The DFBW research program, which spanned 13 years, is considered one of the most significant and most successful NASA aeronautical programs since the inception of the Agency.

Beginning in 1974, NASA Langley's 737 research aircraft flight-tested a variety of large transport aircraft technologies, such as "glass cockpits," airborne wind shear detection, microwave landing systems, and head-up displays. After Langley pioneered the glass cockpit concept in ground simulators and demonstration flights, Boeing developed the first glass cockpits for production airliners. The success of the NASA-led glass cockpit work is reflected in the total acceptance of the electronic flight displays that began with the introduction of the Boeing 767 in 1982. Both airlines and their passengers benefited. Flight safety and efficiency were increased with improved pilot understanding of the airplane's situation relative to its environment.

Ames also led several innovative programs during the 1970s. The Center's Quiet Short-Haul Research Aircraft program developed and demonstrated technologies necessary to support short-takeoff and high-lift cargo aircraft. The aircraft documented stable flight at lift levels three times those generated on conventional aircraft and operated aboard an aircraft carrier without the need for launch catapults or landing arresting gear. Its technologies were employed on the Air Force's C-17 Globemaster II.

NASA's XV-15 tiltrotor research aircraft was the first proof-of-concept vehicle built entirely to Ames' specifications. In 1976, the aircraft hovered for the first time. Two years later, it demonstrated conversion and forward flight as the first tilting rotor vehicle to solve the problems of "prop whirl." Its success directly led to the Marines' V-22 Osprey development, as well as current development of the Bell 609 civil tiltrotor.

In 1979, wing tip "winglets," an invention by Langley's Richard Whitcomb, were introduced to improve vehicle aerodynamics and improve fuel efficiency. Applied at the tips of an aircraft's main wing, winglets are seen on many of today's advanced aircraft, as the technology is now universally accepted. In 1997, The Gulfstream V aircraft incorporated Whitcomb's supercritical wing characteristics and winglets to set 46 world and national performance records. Other aircraft that incorporate these innovations are the Boeing 777 and the C-17.

Remotely Piloted Vehicles

The concept of using remotely piloted vehicles for aeronautical testing and research was first introduced by NASA at the Dryden Flight Research Center in 1969 as a way of eliminating the need for a pilot on a high-risk flight project. Today, remotely piloted aircraft are important engineering tools for aeronautical researchers. Known as remotely piloted research vehicles (RPRVs), these aircraft help NASA improve flight safety, lower flight test and development costs, and improve aircraft construction, materials, and systems.

Between 1979 and 1983, two RPRVs were flown in one of Dryden's most successful research projects. The vehicles, called Highly Maneuverable Aircraft Technology (HiMAT), were utilized to explore and develop high-performance design and structural technologies that could be applied to future aircraft. The rear-mounted swept wings, digital flight control systems, and forward controllable canards gave the vehicles a turn radius twice as tight as conventional fighter aircraft. The RPRV concept was chosen for the program because the experimental technologies and high-risk maneuverability tests would have endangered pilots.

About 30 percent of the aircraft's construction materials were experimental composites such as fiberglass and graphite-epoxy, which allowed them to withstand high-force-of-gravity conditions. Knowledge gained from the HiMAT program strongly influenced other advanced research, and these types of structural materials are now commonly incorporated on commercial and military aircraft. The program produced considerable data on integrated, computerized controls; design features such as aeroelastic tailoring, close-coupled canards, and winglets; the application of new composite materials for internal and external construction; a digital integrated propulsion control system; and the interaction of these then-new technologies upon one another.

The Next Mode of Space Transportation

A new chapter in NASA's history started on January 5, 1972, when President Richard Nixon endorsed plans for the Agency to build a new space vehicle. NASA's Space Shuttle, unlike earlier expendable rockets, was designed to be launched multiple times, serving to ferry payloads and personnel to and from space. As the Space Shuttle concept was being developed, NASA assigned areas of program responsibility to its centers. Kennedy assumed design for ground support facilities and systems for the Shuttle. Johnson led the Shuttle Program and was responsible for the design and procurement of the orbiter. Marshall was tasked with the design and procurement of the external propellant tank, the three main engines of the orbiter, and the solid rocket boosters.

A close-up view of a Space Shuttle Main Engine during a test at Stennis Space Center shows how the engine is rotated to evaluate the performance of its components under simulated flight conditions.

The Highly Maneuverable Aircraft Technology subscale research vehicle, seen here during a research flight, was flown by Dryden Flight Research Center from 1979 to 1983. The aircraft demonstrated advanced fighter technologies that have been used in the development of many modern high performance military aircraft.

In May 1975, the first Space Shuttle Main Engine was tested at Stennis. The main engines that boost the Space Shuttle into low-Earth orbit were flight certified at Stennis on the same test stands used during the Apollo Program. Columbia, the first orbiter scheduled for space flight, was delivered to Kennedy in March 1979, where it began flight processing for its first launch on April 12, 1981. The partially reusable space vehicle was the first flown aerodynamic, winged vehicle to reenter Earth's atmosphere from space, employing technologies developed over 30 years. By the end of 1985, three more orbiters arrived at Kennedy: Challenger, Discovery, and Atlantis.

Aviation Advances in the 1980s

NASA continued to make its mark on civilian flight after the first Space Shuttle mission. When Ames demonstrated a head-up guidance display in a Boeing 727-100 transport airline in the early 1980s, the aviation industry subsequently adapted the technology and certified it for civil transport operations. Riblets, another NASA development, also impacted commercial aviation during this time. Invented by Langley, riblets are small, barely visible grooves that are placed on the surface of airplanes. The V-shaped grooves reduce aerodynamic drag, translating into fuel reductions and significant savings for U.S. commercial airlines.

The Laminar Flow Control project that took place at Dryden between 1986 and 1994 sought to provide similar benefits. By developing active flow control over all speed regimes, the project produced laminar flow over 65 percent of the wing of an aircraft, reducing drag and promoting better fuel efficiency. NASA's research to improve laminar flow dates back to 1930 when NACA photographed airflow turbulence in Langley's Variable Density Tunnel.

From January 1981 through January 1988, nearly 400 commercial airline traction-related accidents occurred as aircraft ran off the ends of runways or veered off shoulders. The resulting crew and passenger fatalities motivated the Landing and Impact Dynamics Branch at NASA Langley to define runway surface maintenance requirements and minimum friction level limits in adverse conditions. Its Safety Grooving research program worked to reduce aircraft tire hydroplaning, the primary cause of uncontrolled skidding during wet weather conditions. The researchers proved that cutting thin grooves across concrete runways to create channels for draining excess water reduces the risk of hydroplaning. As a result, hundreds of commercial airport runways around the world have been safety-grooved. The NASA program improved aircraft tire friction performance in wet conditions by 200 to 300 percent, and countless lives have been saved as a result.

Langley Research Center's Boeing 737 research aircraft is fitted with a Doppler radar wind shear detection system that sends a beam well ahead of the airplane to detect microbursts.

The airborne wind shear detection system that was developed and refined at Langley is another NASA technology that contributes to aircraft landing safety. Wind shear occurs when invisible bodies of air are traveling in different directions of each other at different speeds. Pilots experience severe difficulty in correcting changes in flight path during a wind shear disturbance, particularly while attempting to land. This invisible aviation hazard is so dangerously unpredictable that about 26 aircraft crashed, resulting in over 500 fatalities between 1964 and 1985.

After a Delta Airlines jetliner was brought down by wind shear near Dallas, Texas, in August 1985, it was evident that something had to be done to provide pilots with greater advance warning of wind shear situations. The Federal Aviation Administration (FAA) and NASA Langley combined forces from 1986 to 1993 to develop better wind shear detection capabilities for the airlines and the military. The first challenge was to learn how to model and predict the phenomenon. Langley developed

Wind shear occurs when invisible bodies of air travel in different directions of each other at different speeds. During wind shear disturbances, pilots experience severe difficulty maintaining their flight path, particularly while attempting to land.

the F-factor metric that is now the standard for determining if the airflow ahead of an aircraft is dangerous wind shear.

The next step was to determine what sort of sensor would be the most effective in detecting the wind shear 10 seconds to 1 minute ahead of a flying aircraft. Langley's 737 flying laboratory, NASA 515, flew over 130 missions into severe weather situations, learning how to hunt the invisible hazards 2 to 3 miles ahead of the aircraft. The resulting technological advances have enabled aircraft to read the speed and direction of invisible particles of water vapor or dust in the wind and provide pilots the necessary advance warning of wind shear conditions.

Doppler radar-based systems were developed based on the Langley research and were commercially certified by several companies. The system had its maiden flight on Continental Airlines less than 2 years after the Langley Wind Shear Program declared "mission accomplished" and concluded testing.

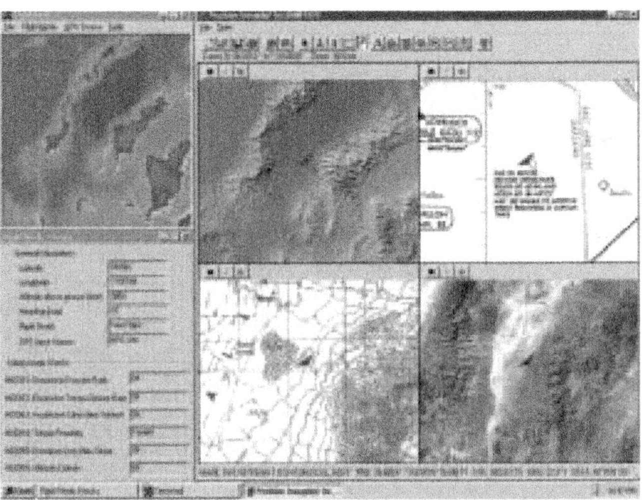

Dubbs & Severino, Inc., creators of the low-cost software packages TerrAvoid and Position Integrity, have designed six warning modes to reflect the FAA categories of concern about safe flight. This image demonstrates how the software packages work together to provide pilots with enhanced situational awareness through the use of six mutually supporting graphic windows.

Safety Strategies in the 1990s

Flying a small plane lost and surrounded by unknown terrain can be a pilot's greatest fear. Through a licensing agreement between JPL and private industry, JPL successfully applied synthetic aperture radar for terrain mapping and Global Positioning Satellite (GPS) data to provide pilots with accurate location and local terrain information in any weather.

In 1994, the Technology Affiliates Program introduced the start-up company of Dubbs & Severino, Inc., to JPL's Dr. Nevin Bryant. Dubbs & Severino had an idea for mapping software to help private airplane pilots, inspired in part by the fatal crash of a pilot friend. The package needed to be completely software-driven, instead of requiring expensive hardware, as was the norm up to that time. Bryant's Cartographic Applications Group at JPL had developed GeoTIFF, an architecture standard providing geolocation tools for mapping applications. GeoTIFF proved to be the crucial key that Dubbs & Severino needed to bring the idea to fruition. With JPL's assistance, the company developed two low-cost software packages that enable pilots to use laptops to detect and avoid hazardous terrain and find their location on maps.

In 1997, NASA created the Aviation Safety Program in response to a report from the White House Commission on Aviation Safety and Security. Forming a partnership, NASA, the FAA, the aviation industry, and the Department of Defense set the program's goal to develop and demonstrate technologies that will contribute to a reduction in the aviation fatal accident rate by a factor of 5 by the year 2007. Langley leads the program, with critical involvement from Ames, Dryden, Goddard, and Glenn. The program's research and technology objectives address accidents involving hazardous weather, controlled flight into terrain, human error-caused accidents and incidents, and mechanical or software malfunctions.

Working within these objectives, the Icing Branch at Glenn supported the development of a new aircraft ice protection system by providing technical and testing support and **Small Business Innovation Research** program funding to Cox & Company, Inc., in 2001. The company's innovation combines an anti-icing system with a mechanical deicer developed by NASA called the Electro-Mechanical Expulsion Deicing System. Together, the two parts form an ice protection system well suited for airfoil leading edges, where ice contamination can degrade aerodynamic abilities. The system has the distinction of being the first aircraft ice protection system to gain FAA approval for use on a new business jet in 40 years.

NASA and Aviation Today

NASA's ongoing research projects contribute to all aspects of aeronautics today. The Agency's emerging technologies have the potential to open a whole new era in aviation, providing advances in air transportation safety and efficiency, national defense, economic growth, and quality of life.

The mission of NASA's Future Flight Central, a fully interactive air traffic control tower simulator located at the Ames Research Center, is to provide a world-class simulation research facility to improve the safety, efficiency, and cost-effectiveness of airport procedures, designs, and technologies. With NASA experts, airport staff can plan new runways, test new ground traffic and tower communications procedures, validate air traffic planning simulations, and perform cost-benefit studies for new airport requirements and designs.

NASA's Intelligent Flight Control, another ongoing research project, developed a neural network technology to help aircraft recover from a loss of control. Current efforts continue to develop neural network technologies that can automatically compensate for damaged or malfunctioning aircraft.

NASA's Environmental Research Aircraft and Sensor Technology (ERAST) project at Dryden serves as a gateway into future aeronautics. Accomplishments such as the solar-powered Unmanned Aerial Vehicle (UAV) Helios prototype, which set an altitude record of 96,863 feet in 2001, are leading the way for future unmanned high-altitude, long-duration, solar-powered aircraft. The ERAST project is also researching a fuel cell-based power system, taking a step toward a UAV that could be sent on missions spanning months at a time.

For over 85 years, the aeronautical contributions of NASA and its predecessor NACA have advanced the safety, efficiency, and cost effectiveness of flight. NASA/NACA technology is on board every U.S. commercial and military aircraft flying today. From wind shear detection and collision avoidance systems to a parachute that lowers an entire aircraft safely to the ground, the aviation benefits derived from the work of NASA/NACA scientists and engineers have impacted the life of every U.S. citizen that has traveled by plane. Through its initiatives and programs, as well as partnerships with the aviation industry, NASA continues to make aeronautics a priority as the United States begins the journey into a second century of flight.

Tests conducted at the Cox & Company Icing Wind Tunnel helped solve the problem of removing ice from airfoils.

NASA's Future Flight Central, the world's first full-scale virtual airport control tower, opened December 13, 1999 at Ames Research Center in Moffett Field, California. The two-story facility is designed to test ways to solve potential air and ground traffic problems at commercial airports under realistic airport conditions and configurations.

The NASA Education Enterprise: Inspiring the Next Generation of Explorers

NASA's challenging and exciting missions provide unique opportunities for engaging and educating the public. Air travel, space flight, the exploration of the unknown, and the discovery of new, mysterious, and beautiful things are all endeavors that hold an intrinsic fascination for people around the world. And the benefits—to NASA, the Nation, and the world—of engaging students and the public in our scientific and engineering adventures cannot be overstated. By stimulating people's imaginations and creativity and by meaningfully communicating the significance of our discoveries and developments to them, we can help to improve the scientific and technological literacy of our society and draw new students into careers in science and engineering.

—NASA 2003 Strategic Plan

Preface

Inspiring our next generation of explorers, inventors, discoverers, technologists, scientists, mathematicians, and engineers is the cornerstone of the new Education Enterprise.

For nearly 50 years, the men and women of NASA have broken barriers to open new horizons of opportunity. Our journeys in air, space, and laboratories have enabled new understanding of our universe, safer and faster air travel, breakthroughs in health care and scientific research, and inspired humanity to reach for new heights. While these achievements and the people behind them are unique, at their foundation they are linked by a common denominator: education. None of the accomplishments we herald in our Nation's history, our daily lives, or in our laboratories and research centers would have been possible without quality education and the people who help open the minds of those who dare to explore and dream.

Educators create horizons of opportunity in classrooms every day. They prepare, inspire, excite, encourage, and nurture the exploration for answers and new questions. Educators are the adventurers in our midst and without them our journeys would not succeed. As we move into our second century of flight, we must work with all who touch the future to help prepare a new generation of Americans to meet the growing challenges in science, technology, engineering, and mathematics.

To meet this challenge, NASA has established the Education Enterprise. Working collaboratively with NASA's scientific and technical enterprises, the Education Enterprise will ensure that education is an integral component of every major NASA research and development mission. This enterprise will provide unique teaching and learning experiences, as only NASA can, through the Agency's research and flight capabilities. Students and educators will work with NASA and university researchers and scientists to use real-time data to study the Earth, explore the universe, and conduct scientific investigations in the fields of aerospace and space-based research.

NASA's Education Enterprise will provide opportunities for students and educators to work with the Agency's scientists and engineers to learn what it takes to develop the new technology required to understand our home planet, explore the universe, and to live and work in space.

As we celebrate the accomplishments of the Nation's first 100 years of flight, we look forward with great anticipation to the next century of flight. The next generation of explorers—the Explorers of the New Millennium—must represent fully this Nation's vibrant and rich diversity. NASA's Education Enterprise will strive to ensure that all children can explore their full potential as Americans. We will fully engage the under-represented and underserved communities of students, educators, and researchers. Furthermore, we will support our Nation's universities, colleges, and community colleges by providing exciting research and internship opportunities that "light the fire" and "fuel the passion" of young people, thereby creating a culture of learning and achievement in science, technology, engineering, and mathematics.

Welcome to NASA's Education Enterprise. Working together, we can "see learning in a whole new light."

Adena Williams Loston

Dr. Adena Williams Loston
Associate Administrator for Education
National Aeronautics and Space Administration

The NASA Education Enterprise: Inspiring the Next Generation of Explorers

"Today, America has a serious shortage of young people entering the fields of mathematics and science. This critical part of NASA's mission is to inspire the next generation of explorers so that our work can go on. This educational mandate is an imperative."

On April 12, 2002, NASA Administrator Sean O'Keefe opened a new window to the future of space exploration with these words in his "Pioneering the Future" address. Thus began the conceptual framework for structuring the new Education Enterprise.

The Agency's mission is to understand and protect our home planet; to explore the universe in search for life; and to inspire the next generation of explorers … as only NASA can. In adopting this mission, education became a core element and is now a vital part of every major NASA research and development mission.

NASA's call to "inspire the next generation of explorers" is now resounding throughout the NASA community and schools of all levels all around the country. The goal is to capture student interest, nurture their natural curiosities, and intrigue their minds with new and exciting scientific research; as well as to provide educators with the creative tools they need to improve America's scientific literacy. The future of NASA begins with America's youngest scholars. According to Administrator O'Keefe's address, if NASA does not motivate the youngest generation now, "there is little prospect this generation will choose to pursue scientific disciplines later."

Since embracing Administrator O'Keefe's educational mandate over a year ago, NASA has been fully devoted to broadening its roadmap to motivation. The efforts have generated a whole new showcase of thought-provoking and fun learning opportunities, through printed material, Web sites and Webcasts, robotics, rocketry, aerospace design contests, and various other resources … as only NASA can.

Administrator O'Keefe selected a highly experienced educator and education administrator, Dr. Adena Williams Loston, in the fall of 2002, to lead the charge for NASA's Education Enterprise. Dr. Loston's mandate from Administrator O'Keefe is to stimulate interest among students in science, technology, engineering, and mathematics study and careers by raising public awareness among educators, students, and parents about the vast array of NASA education programs and resources available. The mandate also calls for the engagement of these individuals in interactive educational activities that highlight the discoveries and missions of NASA's Offices of Space Flight, Space Science, Aerospace Technology, Earth Science, and Biological and Physical Research.

In 2002 alone, NASA reached well over half a million educators, nearly two million students in grades K through 12, and almost 70,000 higher education students through direct, on-site activities and programs. In addition to those served by broad-based NASA Education programs, the Agency also directly reached over 17,000 minority students through its minority-targeted academies, scholarships, and initiatives.

Whirlybirds, Pigeons, and Bats, Oh My!

In one particular endeavor to inspire the Nation's youth, NASA decided in 2002 to build on children's fascination with flying vehicles. "Robin Whirlybird on Her Rotorcraft Adventures," an online, interactive children's book, is introducing kindergarteners through fourth graders to the history, concepts, and research behind aeronautics and rotorcraft. Designed to have the look and feel of a children's book, the story revolves around a young girl named Robin who visits a NASA research center where her mother works as an engineer. During her visit, Robin explores the concepts of aeronautical design, the physics of flight, and the practical application of rotorcraft, also known as helicopters or runway-independent aircraft. The book incorporates interactive elements on every page, a menu bar with various exploration buttons, and lesson plans aimed at strengthening language arts and vocabulary skills, making it a unique classroom tool.

In a similar effort also launched last year, NASA's "The Adventures of Amelia the Pigeon" project is teaching children how scientists use satellite imagery to better

Designed to have the look and feel of a children's book, "Robin Whirlybird on Her Rotorcraft Adventures" revolves around a young girl's visit to a NASA research center where her mother works as an engineer.

understand the Earth's environmental changes. The Web site acquaints young students (grades K through four) with Earth science concepts, beginning with classifying objects by shape, color, and texture, building a foundation for interpretation and understanding of remote sensing. The pigeon adventure—based in New York City, because of its size, diversity, and visibility of prominent features in satellite imagery—encourages the development of children's inquiry skills, via online explorations, sequential storytelling, and hands-on investigations. The story also features supplemental classroom materials associated with National Science Education Standards. "Amelia the Pigeon" is a result of Goddard Space Flight Center's Interactive Multimedia Adventures for Grade School Education Using Remote Sensing (IMAGERS) collaboration with the Department of Interior's United States Geological Survey, and follows on the success of the "Echo the Bat" program, which teaches children to understand light and the electromagnetic spectrum, as they apply to remote sensing.

Students in the fourth grade at Longfellow Elementary School, Columbia, Maryland, answer questions about the solar system, using a planetary model called an "orrery" that they painted after viewing Hubble Space Telescope pictures.

New Series of Science Books

In March of this year, NASA and Pearson Scott Foresman, the leading pre-K through sixth grade education publisher, formally agreed to collaborate on elementary and middle school science curricula. Under the terms of the partnership, Pearson Scott Foresman editors and authors will draw upon NASA's rich archival material and extensive research in biological, physical, Earth, and space sciences to create the Scott Foresman Science series of books. NASA experts will review the content, and Pearson Scott Foresman will ensure the curricula reflect National Science Education Standards, as well as other specific targeted standards. Before the new series is published, specific lessons will be developed for students and teachers, following the steps that Barbara Morgan, NASA's first Educator Astronaut and a second and third grade teacher, is taking in preparation for flight into space.

Astronomy

Further helping students to reach for the stars, a new NASA astronomy program is bringing together existing Internet technology and other tools to open the universe to those who would otherwise be denied the experience due to their physical or cognitive disabilities. The effort, funded by NASA through the Space Telescope Science Institute (STScI) of Baltimore, involves the participation of the elementary school system in Howard County, Maryland. Dr. Carol Grady, a National Optical Astronomy Observatory researcher stationed at Goddard Space Flight Center, is the science lead for the program. She became involved after her son, who has special needs, expressed an interest in her work with the Hubble Space Telescope on planet formation and stellar evolution. "The advances in astronomy over the last hundred years are one of humanity's greatest cultural achievements, and I did not want kids like my son to get the message that activities like this are not open to them," said Grady. The team chose to target elementary-age students so that it can deliver assistive technology to them before frustration leads them to give up attempting to learn.

The wonders of the universe are also being brought to the fingertips of visually impaired students in a new book titled "Touch the Universe: A NASA Braille Book of Astronomy." The 64-page book presents color images of planets, nebulae, stars, and galaxies. Each image is embossed with lines, bumps, and other textures. The raised patterns translate colors, shapes, and other intricate details of the cosmic objects, allowing the visually impaired to feel what they cannot see. The book incorporates Braille and large-print descriptions for each of its 14 photographs to make it accessible to readers of most visual acuities. "Touch the Universe" takes the reader on a cosmic journey. It begins with an image of the Hubble Space Telescope orbiting Earth, and then travels outward into the universe, showing objects such as Jupiter, the Ring Nebula, and the Hubble Deep Star Field North. The author, Noreen Grice, teamed with

Bernhard Beck-Winchatz, an astronomer at DePaul University in Chicago, to develop the book with a $10,000 Hubble Space Telescope grant for educational outreach. Students at the Colorado School for the Deaf and Blind in Colorado Springs evaluated the early prototype images for clarity and provided suggestions for improvement prior to publication.

Space Day 2003

Students of all ages gathered at various sites around the world on May 1 to pay tribute to aerospace exploration and to celebrate the "Future of Flight." Administrator O'Keefe and Senator John Glenn kicked off the celebration with an opening ceremony at the Smithsonian's National Air and Space Museum in Washington, DC. During the ceremony, student teams and teachers around the country were recognized for their "stellar" future spacecraft designs. The "Young Ohio Engineers" from Grace Home School in Westerville, Ohio, was the team awarded with the best overall "Fly to the Future" design among fourth and fifth graders. This five-student lineup designed a multifunctional X-76 Independence aircraft that could be used for various tasks, including commercial transport and military use. The X-76 Independence contains many interchangeable parts (based on its mission), is powered by a combination of turbo jet and scram jet engines that change modes based on altitude and speed, and has a vertical takeoff and landing system that enables it to be used almost anywhere.

In the "Planetary Explorers" challenge, "Team Jupiter" from the Franklin Magnet Middle School in Champaign, Illinois, took home the best overall award among sixth through eighth graders for their creation. Team Jupiter decided to go to Europa, one of the moons of Jupiter, and conduct scientific experiments. Their "CSSC-BAM V" vehicle featured a blended-wing body design made from titanium alloy. It would use jet engines when flying on Earth or to near-Earth orbit, and plasma engines in space. A team of snakebots—machines that can crawl, coil, climb, and grasp, just like serpents—each with its own unique task, manned the plane to explore the surface of Europa upon arrival.

The celebration continued around the globe. In Morris Plains, New Jersey, first graders at Mountview Road School attempted to simulate dockings, manipulate tools under water, and eat food as if they were in a weightless environment. In Crawfordville, Florida, students viewed space videos, built small LEGO machines using NASA glove boxes they made, read space exploration books, and ate moon pies. At the Museum of Science in Boston, Massachusetts, Jet Propulsion Laboratory Solar System Ambassador Charlie Haffey spoke about the exploration of Mars, Jupiter, and Saturn, which prompted students to construct scale models of the planets. Internationally, students from Arecibo, Puerto Rico, toured the world's largest radio telescope that searches for life in distant galaxies. In Pukekohe, South Auckland, New Zealand, students were treated to space awareness workshops featuring hands-on activities.

Over 500 schools worldwide participated in the "Student Signatures in Space" program to sign posters that will be digitized and eventually flown on a Space Shuttle mission. Established in 1997, Space Day is dedicated to the extraordinary achievements, benefits, and opportunities in the exploration and use of space. NASA is one of more than 75 organizations that support the award-winning educational initiative.

SEMAA: A Decade of Devotion to Learning

Over the last 10 years, an innovative program managed by the Office of Educational Programs at NASA's Glenn Research Center has been inspiring a diverse student population in grades K through 12 to pursue careers in the fields of science, engineering, mathematics, and technology. The 10th anniversary celebration of the

Administrator Sean O'Keefe visits with students at Space Day 2003—Celebrating the Future of Flight. Space Day, the annual tribute to aerospace exploration, invited young students to honor the previous 100 years of aviation accomplishments at the Smithsonian's National Air and Space Museum on May 1.
Photo credit NASA/Renee Bouchard

Science, Engineering, Mathematics, and Aerospace Academy (SEMAA) coincided with the annual National SEMAA Conference held in June in Cleveland, Ohio. Educators, students, parents, and administrators from all 19 SEMAA sites across the country attended. The focus of the conference was to develop additional ways to create awareness and access to programs with similar purposes and to develop partnerships. SEMAA was born in 1993 of then U.S. Congressman Louis Stokes' concern about the low level of academic achievement of the young students in his district and his urging for the creation of a unique program that would focus on mathematics and science.

NASA Explorer Schools

A new NASA education initiative has been designed to provide customized, extended professional development for educators and unique NASA science and technology learning experiences for students. The 3-year NASA Explorer School (NES) program will align participating schools with NASA personnel and other partners to develop and implement action plans for teachers and administrators. The action plans will promote and support the use of NASA materials and programs that address local needs in mathematics, science, and technology. Fifty NES teams will be chosen from around the country. Each team will consist of three or four science, mathematics, or technology educators, an administrator, and a state supervisor, and will participate in an expense-paid week of intensive training at one of NASA's 10 centers. Each team will also receive a $10,000 grant, intended to assist with the purchase of science and technology tools to support implementation plans and bring cutting-edge technology to the classroom. The 2003 pilot year focus is for grades five through eight.

Emmy Award-Winning Television Programs

In an effort to continue promotion of higher learning through educational television programming, NASA is joining forces with South Carolina Educational Television (SCETV) to video-stream three educational television series to classrooms throughout South Carolina and across all other states. Developed by Langley Research Center's Office of Education, the Emmy Award-winning shows "NASA Science Files" and "NASA Connect" are aimed at students ranging from grades 3 through 12; the third show, "NASA's Destination Tomorrow," is also an Emmy Award recipient, but is designed for educators, parents, and life-long learners. NASA is using SCETV facilities to broadcast the three series nationwide. Approximately 18,000 South Carolina educators, representing about 500,000 students, are registered users of the programs.

Webcasts

NASA Webcasts are also becoming more prevalent in America's schools. Ames Research Center's "NASA Quest" Web site, a rich resource for educators, children, and space enthusiasts interested in meeting and learning about the people who work for the Space Program, features a full calendar of audio/video Webcasts and live, interactive events. On any given day, students and others can log on and learn why a deep-sea submersible laboratory stationed in the Florida Keys is helping NASA astronauts to prepare for long-term space travel; why NASA is studying the Northern Lights phenomena to improve satellite operations and space communications; and how a catalytic carburetor designed by NASA will help to reduce air pollution. NASA Quest also connects schools with NASA staff through Web chats, forums, e-mail, informative biographies and journals, curriculum resources, and more.

In March, students from grades 5 through 12 explored the frozen landscapes of Colorado's Rocky Mountains—from their desks. As virtual participants in two live Webcasts, students nationwide joined scientists from NASA, the National Oceanic and Atmospheric Administration (NOAA), other Federal agencies, and many universities as they studied the role of snow-cover on the Earth's water and climate. Using skis, snowmobiles, aircraft, and satellites, scientists participating in the 2003 NASA-NOAA Cold Land Processes Experiment studied snowpack from the ground, air, and space across the winter and spring of this year to improve forecasts of springtime water supply and snowmelt floods. Through interaction, students gained an understanding of how remote sensing is used in Earth science research and how

The NASA Connect team visited students at Andoya High School in Andenes, Norway, to tape an instructional video as part of its annual series of free standards-based learning programs.

scientists verify data from airborne platforms and satellites hundreds of miles above the planet.

Frozen ice was also the topic of a series of live Webcasts in April. Secondary and college classrooms were invited to participate in the Internet broadcasts to explore the frozen ice sheets of the North Pole and learn how they play a role in warming the Earth. The actual research was carried out by NASA scientists and Native American students from the Bay Mills Community College in Brimley, Michigan. Together, the scientists and college students gathered data about the nature and thickness of sea ice in the Arctic and measured the concentration of aerosols and their specific properties, such as size and absorption of sunlight (the amount of sunlight that aerosols absorb is important in helping scientists better understand how they contribute to trapping heat in the atmosphere and warming the Earth). The purpose of involving the Bay Mills students was to inspire Native American students to seek out careers in technology and science. Students watching the Webcasts from their classrooms had the opportunity to dialogue with the scientists and the students in the field.

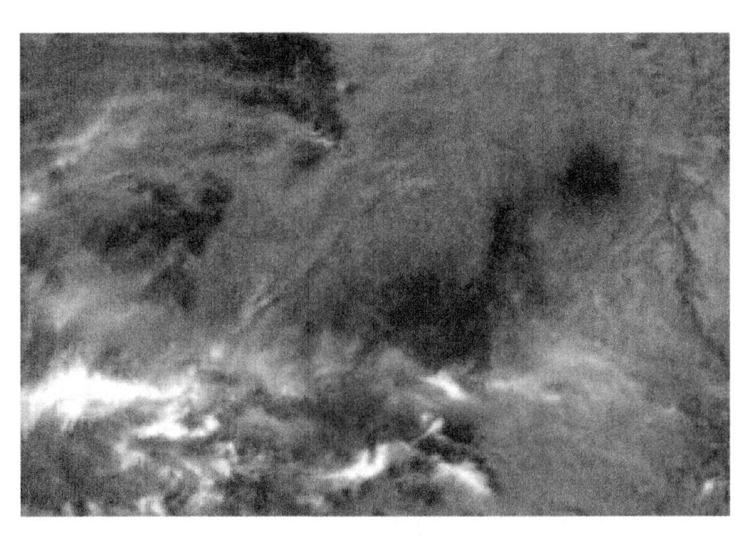

An aerial image of Arizona and Utah, taken from the International Space Station by middle school students, via Internet connections.

In May, an interactive Webcast gave students an early look at NASA's plans to land two twin robotic geologists on Mars in January 2004. The hour-long "Countdown to Mars" program, hosted by "Bill Nye the Science Guy," invited 250 students to conduct science and engineering experiments based on those of the actual Mars Exploration Rover mission. Viewers throughout North America were able to interact via e-mail as the students carried out the experiments on camera. Jet Propulsion Laboratory's Dr. Joy Crisp, the rovers' project scientist, joined the program as one of its guests.

Another Webcast that took place in May offered eighth graders from economically disadvantaged Chicago-area schools the opportunity to see science in action and be inspired by the International Space Station crew. The event, hosted by Chicago's Adler Planetarium and Astronomy Museum, "connected" the students with Expedition Seven astronaut Ed Lu and cosmonaut Yuri Malenchenko. The Space Station Webcast was just one of many enabled by NASA's Teaching in Space Program, managed by Johnson Space Center.

Snapshots From Space

In a case of long-distance learning, middle school students from McNair Magnet School in Cocoa, Florida, were the latest participants in a NASA project that allows youngsters to take pictures of the Earth using a camera on the International Space Station (ISS). The ISS EarthKAM (Earth Knowledge Acquired by Middle school students) project, created in 1994 by Dr. Sally Ride—America's first woman astronaut, is helping scientists from Goddard Space Flight Center study the planet's changing surface. From April 29 to May 2, the McNair Magnet students controlled the high-resolution digital camera operating on the Space Station's Destiny science module, via Internet connections. The students profited by being involved in the process of real scientific research and through their interaction with scientists as they worked together on the analysis of research images. The next picture-taking mission is scheduled for November 2003, followed by four more in 2004.

Contests to Challenge the Mind

The two twin robotic geologists being sent to Mars now bear the names "Spirit" and "Opportunity," thanks to a 9-year-old explorer-to-be. Sofi Collis, a third grader from Scottsdale, Arizona, wrote the winning essay in a contest to name the rovers. Collis' essay was selected from nearly 10,000 entries in the contest, sponsored by NASA and Denmark-based toymaker LEGO, Co., with collaboration from the Planetary Society of Pasadena, California. Collis, who was born in Siberia and brought to the United States through adoption, read her essay at the name-unveiling ceremony in June at Kennedy Space Center: "I used to live in an orphanage. It was dark and cold and lonely. At night, I looked up at the sparkly sky and felt better. I dreamed I could fly there. In America, I can make all my dreams come true. Thank you for the 'Spirit' and the 'Opportunity.'"

Explorer-to-be Sofi Collis with a Mars Exploration Rover model.

Nobel laureate physicist Luis Alvarez and his coworkers proposed in 1980 a theory that an asteroid collided with the Earth about 65 million years ago, causing the extinction of the dinosaurs. The Earth has experienced catastrophic changes in its history that have changed the course of biological evolution many times. Should disaster strike the Earth, a large space colony civilization can insure life's survival and, if possible, succor Earth, according to NASA scientists. The idea of space colonies is also a natural curiosity, because living things want to grow and expand, like weeds that grow through cracks in sidewalks, and living creatures that crawl out of the oceans and colonize land. NASA notes that the key advantage of space settlements is the ability to build new land, rather than take it from someone else. This allows, but does not guarantee, a huge expansion of humanity without war or destruction of Earth's biosphere.

NASA hosts an annual Space Settlement Contest, sponsored by the Fundamental Space Biology Program at Ames Research Center, to encourage students to develop the ideas and skills necessary to make orbital colonies a possibility. The contest challenges students in grades 6 through 12 to investigate and then develop designs for a permanent, relatively self-sufficient home that cannot be based on a planet or a moon. The Fundamental Space Biology Program created a Web site that provides students access to a wealth of electronic resources to help in developing designs. The 2003 grand prize winners were two middle school students from Romania. Horia Mihail Teodorescu and Lucian Gabriel Bahrin submitted the design for an orbital colony called "Teba 1." The winning design was chosen by a panel of NASA scientists from a field of 89 designs submitted by 307 students from the United States, Austria, India, Japan, and Romania. The grand prize winners, along with the first-, second-, and third-place winners in the individual and small group categories, were invited to Ames to present their designs, talk to NASA scientists, and tour the Fundamental Space Biology laboratories.

More creative problem-solving took place during the 26th annual "Odyssey of the Mind World Finals," held at Iowa State University in May. Sponsored by NASA's Earth Science Enterprise, the Odyssey contest tested the abilities of students of all ages from around the world to design, construct, and run three small vehicles to transport items from an orbit area to an assembly station in "space." The NASA problem opened with a three-dimensional representation of an Earth scene as viewed from space. Items affecting the problem, both real and imaginary, were added to the scenario, effecting a scene change. The vehicles were powered in different ways: one carried its energy source, and two traveled on the momentum created by different sources. This problem was one of five long-term challenges presented at the contest. During the year, students were separated into four divisions, based on age, and formed teams to solve one of the challenges. After developing their solutions, teams competed at state and regional levels, before moving on to the World Finals.

Students are also receiving an educational "boost" through a variety of rocketry challenges. More than 1,000 students from 100 high schools throughout the United States gathered in Virginia to compete in the inaugural Team America Rocketry Challenge, considered the world's largest model rocket contest. The event, held in conjunction with the national yearlong

Image courtesy of The LEGO Company

A LEGO 1:1 scale model of the Mars Exploration Rover, shown on display at the World Space Congress in Houston, Texas, in October 2002. The rover model weighs 130 kilograms (290 pounds), is made from approximately 90,000 LEGO elements, and took 650 man-hours to build.

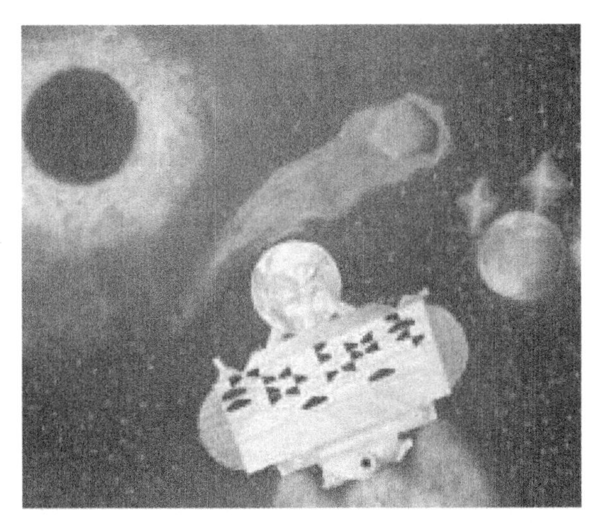

Space Settlement Contest: Design submitted by students from Kadena Middle School in Okinawa, Japan (2002).

Centennial of Flight Celebration, offered student teams awards worth $59,000. Boonsboro High School of Boonsboro, Maryland, placed first in the contest. The top 10 teams became eligible to submit proposals to participate in the 2004 Student Launch Initiative at Marshall Space Flight Center, where students build reusable launch vehicles carrying a science experiment payload up to an altitude of 1 mile.

Elementary and high school students from five states worked feverishly for 9 months across 2002 and 2003 to prepare their experiments for launch aboard a NASA rocket to the upper limits of the Earth's atmosphere. In June, they saw their hard work pay off with the launch of the 20-foot-tall NASA sounding rocket from the Wallops Flight Facility in Virginia. The high school students' experiments focused on satellite communications, spectral imaging and analysis, and materials and fluids in a high-stress environment, while the elementary students' experiments focused on static electricity. The flight, part of the NASA Student Involvement Program, exposed the experiments to stresses 15 times Earth's gravity.

In the realm of robotics, hundreds of students from the United States and Canada converged on Cleveland, Ohio, in March for the Buckeye Regional For Inspiration and Recognition of Science and Technology (FIRST) Robotics Competition. Working side-by-side with professional engineers and technicians, the students took a hands-on approach to discover what real-world engineering is all about. The students were divided into teams for a game called "Stack Attack," in which they maneuvered robots they built to detect and attack opposing robots' stacks of plastic containers. Participating robots were required to operate autonomously using onboard sensors to seek out the containers. The students would then take control, commanding their robots to position as many containers as they could on their side of the playing field and also stack them as high as possible. Although each team started out with the same parts kit to build its robot back in January, 64 unique robotic creations were represented at the competition.

In another region, students gathered at Houston, Texas' Reliant Park for the FIRST Robotics Lone Star Competition. They too, participated in a game of Stack Attack to determine the best functioning robots. The winners of the Cleveland and the Houston competitions—sponsored by Glenn Research Center and Johnson Research Center, respectively, in cooperation with local corporations, educational institutions, and organizations—and 21 other regional U.S. contests competed in the FIRST Robotics Championship Competition in May. On a whole, NASA and its Robotic Education Project sponsored 207 of the nearly 800 teams entered in the 2003 FIRST competition. Regional and national awards were presented to students for excellence in design,

At the 2003 For Inspiration and Recognition of Science and Technology (FIRST) Robotics Competition, students participated in a game called "Stack Attack."

engineering innovation, control systems, demonstrated team spirit, sportsmanship, and many other categories.

NASA's Offices of Space Flight and Aerospace Technology, through the NASA Quest program, sponsored a separate robot-design initiative called the Robotic Helper Design Challenge. In May, the 2-month educational activity brought students and NASA experts together for a live Webcast to review the progress of the students' designs, intended to help astronauts living and working on the ISS. Through the Webcast, the students engaged in a virtual tour of the ISS, learned about microgravity, and had questions about their designs answered by NASA experts. An estimated 2,500 students in 100 classrooms (representing 26 states and 7 countries) took part in the final design challenge. Entries included a Space Pet Involving Kinetic Energy (SPIKE) robot that uses propellers to steer itself while floating through low-gravity atmospheres (created by eighth graders at Barkalow Middle School in Freehold, New Jersey), and "Mr. Helper," which comes equipped with a homing device so that it can return to its docking bay to recharge, and a large storage compartment for astronauts' tools (created by fourth, fifth, and sixth grade students at the K.R. Smith Elementary School in San Jose, California). The inspiration for the design challenge was NASA's prototype Personal Satellite Assistant, an astronaut-support tool devised to move and operate autonomously or by remote control in the microgravity environment of the Space Shuttle, ISS, or a future space vehicle.

Collegiate Research Opportunities

Students closing in on their undergraduate and post-graduate degrees are really getting a taste of what it is like to work for NASA, through real-life research opportunities. At North Carolina State University, students enrolled in an aerospace design class are helping NASA expand the exploration of Mars' surface. The team of students and researchers designed a wind-powered rover that can be blown, like a tumbleweed, across the surface of the Red Planet, for the purpose of collecting atmospheric geological samples. To create the Tumbleweed Earth Demonstrator rover, the team studied Langley Research Center concepts, researched wind tunnel testing, and performed actual field-testing. The student-built rover is expected to provide preliminary data that will influence future tumbleweed design concepts.

Three students working toward graduation and advanced degrees are looking forward to adding the title of "inventor" to their names. As participants in NASA's Undergraduate Student Research Program through Marshall Space Flight Center, Amanda LaZar, Dave Broderick, and Andrew Schnell have suggested innovations that soon could be used by the Space Program to increase safety, facilitate inspection and maintenance of delicate equipment, and create lightweight structures strong enough to withstand the harsh environment of Earth orbit.

A team of North Carolina State University students designed a wind-powered rover that can be blown, like a tumbleweed, across Mars' surface, to collect atmospheric geological samples.

LaZar, a mechanical engineering student at the University of South Carolina on pace to graduate in 2003, anticipates receiving a patent for finding a way to weld joints on the Space Shuttle External Tank that will both improve safety and reduce repair costs. Broderick, an electrical engineering and computer science student at Hartford University also on course to graduate in 2003, completed his first summer as an undergraduate researcher at Marshall in 2002. He helped develop a vision-based guidance system for a miniature robot, allowing technicians to make inspections and repairs without having to dismantle the apparatus. Schnell, now in his third year of the student research program after having graduated magna cum laude from the Georgia Institute of Technology in 2002, came up with a new manufacturing process that uses balloon-like material inflated with gas and filled with hardened foam to create beams and other structures. Expected to be patented, it has potential for both space and ground uses, such as space solar power systems or sporting equipment. If used in place of conventional space structure materials such as metal alloys, Schnell's innovation could drastically cut payload weights on the Space Shuttle, which currently cost about $10,000 per pound to launch.

Each year, the Undergraduate Student Research Program offers undergraduates across the Nation mentored research experiences at participating NASA centers, through fall and summer sessions. NASA additionally hosts the Graduate Student Researchers Program to award fellowships for graduate study leading to research-based masters or doctoral degrees in the fields of science, mathematics, and engineering.

Six students from the University of New Mexico, Oregon State University, Utah State University, and the University of Utah are spending the summer of 2003 monitoring the West Nile virus, studying satellite images to assess the potential for dangerous wildfires, and embarking on many other educational adventures involving natural resources. Selected to receive training and complete internships in applied Earth science under the "Develop" program, the students will lead investigations by applying NASA technology to local concerns. Develop provides workforce growth and outreach to communities, enabling students to tap science to help solve real-world problems. The 10-week project began in June at Ames Research Center. The primary objective of the West Nile virus study is to identify potential mosquito habitats and correlate the data to human populations at high-risk. For the wildfire task, the students are mapping and monitoring invasive and noxious plant species that act as fuel for wildfires.

Partnerships to Encourage and Inspire

During a week in May, NASA joined Career Opportunities for Students with Disabilities (COSD) at the group's annual meeting in Redmond, Washington. COSD is actively helping NASA find qualified students with disabilities who are pursuing mathematics, science, engineering, and technical degrees for employment with the Agency. NASA also supports COSD by providing outreach and assistance to Historically Black Colleges and Universities and Minority Institutions to encourage recruitment, development, and academic growth opportunities.

In the past year, NASA helped to launch a new education center to inspire and support socially and economically disadvantaged students in their quest for higher learning. The NASA Center for Success in Math and Science, located on the Avondale, Arizona, campus of Estrella Mountain Community College, was dedicated by NASA astronaut Carlos Noriega and Estrella President Homero Lopez. The center not only reaches out to local Hispanic students in metropolitan Phoenix, it also engages Hispanic-Serving Institutions currently not involved in NASA programs, especially community colleges. Through the center, NASA will provide educators with unique resources to create learning opportunities that support educational excellence, encourage family involvement, and establish links with local business and community groups.

Astronauts Barbara Morgan and Leland Melvin visit with students and teachers at the Hardy Middle School in Washington, DC.

NASA recently launched its 2003 Summer High School Apprenticeship Research Program (SHARP) after competitively selecting 340 high-achieving students representing nearly every state in the Nation and the U.S. territories of Puerto Rico and St. Croix. In June, NASA SHARP participants, chosen from a pool of more than 2,400 applicants, became apprentices to scientists and engineers at NASA centers and universities around the country. NASA SHARP is a synergistic, research-based program that focuses on NASA's mission, facilities, human resources, and other programs. The effort advances the Agency's goal to involve underrepresented students in academic, workplace, and social experiences, as well as research opportunities to support the educational excellence of the Nation.

In another partnership, NASA and the Foothill-De Anza Community College District will facilitate the development of an academic center in NASA Research Park at Ames Research Center for first-generation college students interested in science, technology, and engineering careers. The agreement will bring community college students to classrooms and laboratories onsite at Ames. Since 1957, the Foothill-De Anza Community College District has responded to the needs of more than 1 million Silicon Valley students.

Educator Astronaut and Earth Crew Programs

It is official: Teachers are more interested in space than ever! NASA's mailroom overflowed with applications from teachers who want to become members of the permanent Astronaut Corps. NASA received over 8,800 teacher nominations during the 3-month recruitment phase; even more, the Educator Astronaut Program office received over 1,600 applications. NASA will review the applications and select Educator Astronaut candidates to begin training with the Astronaut Corps at Johnson Space Center. After graduation, new Educator Astronauts will be eligible for a Space Shuttle flight assignment as fully trained Mission Specialists.

To promote the program and encourage students to nominate their teachers, astronaut Barbara Morgan and Educator Astronaut Program co-managers Debbie Brown and Leland Melvin (also an astronaut) visited many schools and organizations around the United States and in Puerto Rico. Melvin, for example, enlightened teachers and students at schools, conferences, and community centers such as Rome Free Academy, Proctor High School, and the Gloria Wise Boys & Girls Club & Community Center in New York; and Elliot Middle School and the Vanguard Learning Center of Compton in California.

As a former National Football League player, Melvin was on his way to stardom when his football career was cut short by an injury. He decided to pursue an alternative passion, engineering, which led to his acceptance into NASA's permanent Astronaut Corps in 1998. He is taking his story on the road to inspire students to follow their dreams and always have a back-up plan. "To accomplish great things, you must not only dream, but also plan; and every plan should contain options, like having a spare tire, just in case you get a flat," Melvin said.

A recent survey conducted by the National Science Teachers Association indicated that more than 91 percent of science teachers should have a place aboard future Space Shuttle flights, to bring the educational value of the missions to the classroom. Science teachers also believe Educator Astronauts could spark student interest in science and mathematics careers, and serve as role models to instill in students how these studies apply to the real world.

Meanwhile, the next phase of engaging students, teachers, and parents to explore space took flight through NASA's Earth Crew activity. International participants are welcomed and encouraged to join the Earth Crew, which currently consists of more than 23,000 U.S. and international members. The Web-based educational program features activities that enable students, educators, and parents to interact with astronauts, scientists, and engineers in projects and missions. New inspiring and educational Earth Crew Missions became available on the Educator Astronaut Web site in May.

With a charter like no other, NASA has led some of the most unique missions in the world. From traveling to low-Earth orbit and walking on the Moon, to viewing the farthest reaches of our solar system, NASA has continually worked to share the discovery and adventure along the way. Each of these achievements is something that only NASA can do; therefore, the Agency is striving even more to share these experiences with inquisitive minds in order to inspire and prepare them for future challenges.

NASA-sponsored education programs create a pipeline that engages a diversity of students in the earliest grades and encourages them to continue through college, graduate school, and postgraduate studies in science, mathematics, engineering, technology, and geography. NASA continues to develop and structure programs incorporating its resources and technologies to inspire the next generation of explorers … as only NASA can.

Making way for future explorers: (from left to right) Dr. Clifford W. Houston, NASA Deputy Associate Administrator for Education Programs; Peggy Steffen, Program Manager, NASA Explorer Schools Program; Dr. Shelley Canright, NASA Director of the Office of Technology & Products, Education Enterprise; and Dr. Huan Ngo, Team Lead, Sheridan Magnet Middle School, New Haven, Connecticut (NASA Explorer School). NASA astronaut Ed Lu (left background) and Russian cosmonaut Yuri Malenchenko (right background) join the NASA Explorer School officials via a live feed from the International Space Station.

Partnership Successes

NASA cultivates partnerships with private industry, academia, and other government agencies to address the challenges that face our Nation. By contributing time, facilities, and technical expertise, NASA brings the benefits of space down to Earth where it enriches the lives of the American public. The following pages are illustrative of the wealth of NASA success stories generated each year.

Partnership Successes

Each year, NASA research and development efforts contribute to a variety of successes. Through partnerships with industry and academia, NASA's space-age technology improves all aspects of society. While not every technology transfer activity results in commercialization, these partnerships offer far-reaching benefits to U.S. citizens. The following examples are just a few of the ways NASA is applying its technology and resources to improve the quality of life on Earth.

Traffic Safety

This year, NASA and the National Highway Traffic Safety Administration (NHTSA) joined forces to, literally, take vehicles out for a spin. The NHTSA currently employs a consumer rating system that uses an engineering formula to determine rollover resistance, but wanted to research new methods through NASA. Goddard Space Flight Center's High Capacity Centrifuge, used to test spacecraft before they are sent into space, was exactly what was needed to spin up some unique and original vehicle testing. Vehicles were positioned on the High Capacity Centrifuge's test platform, and spun until inertia and centrifugal force caused them to tip. A crash-test dummy was along for the ride in each vehicle to increase the realism and accuracy of the test results. The High Capacity Centrifuge is a big machine, more than 150 feet in diameter, filling an entire circular building. With two powerful motors running at full tilt, the outer edge of the test arm can reach speeds of more than 200 miles per hour, producing a force 30 times Earth's gravity. At rest, the giant multi-ton arm sits on bearings so smooth just two or three people can push it around the room. NASA and the NHTSA expect this first-of-its-kind test will enable them to gain valuable safety information about vehicles that move millions of Americans every day.

Environmental Conservation

In an effort to preserve Earth's natural resources, Goddard is the first Federal facility to heat its buildings with landfill gas. By harnessing methane gas from a nearby landfill and utilizing it to fire steam-producing boilers, Goddard is reducing emissions equivalent to taking 35,000 cars off the road per year, or planting 47,000 acres of trees, according to Barry Green, the Center's Energy Manager. On top of this, officials claim NASA will save taxpayers more than $3.5 million over the next decade in fuel costs. A few years ago, Dallas-based Toro Energy, Inc., approached NASA, offering landfill gas as a way to reduce fuel costs while helping to protect the environment. At no cost to the government, Toro Energy built a purification plant and a 5-mile pipeline from the Prince George's County, Maryland-based Sandy Hill Landfill to Goddard, and modified two boilers at the Center. The Sandy Hill Landfill has collected about 5.2 million tons of trash and is expected to generate landfill gas for at least 30 years; NASA plans to use the gas for 10 to 20 years. The Environmental Landfill Agency's Landfill Methane Outreach Program also provided expertise to help complete this project.

Earth Science

Scientists at Goddard, the Jet Propulsion Laboratory (JPL), and Ames Research Center are working in conjunction with several universities to develop an advanced earthquake modeling system. QuakeSim will give researchers new insight into the physics of earthquakes using state-of-the-art modeling, data manipulation, and pattern recognition technologies when it is completed in 2004. Consisting of several simulation tools, QuakeSim will generate new quake models that researchers anticipate will vastly improve future earthquake forecasting. According to QuakeSim principal investigator Dr. Andrea Donnellan, from JPL, the forecasts can be used by a variety of Federal and State agencies to develop decision support tools and help mitigate losses from large earthquakes.

Emergency Management

A new emergency communication system that aids first responders to natural or

A sports utility vehicle harnessed to NASA's High Capacity Centrifuge is being prepared for a spin to simulate the kinds of forces that might cause the vehicle to lose its stability and roll over.

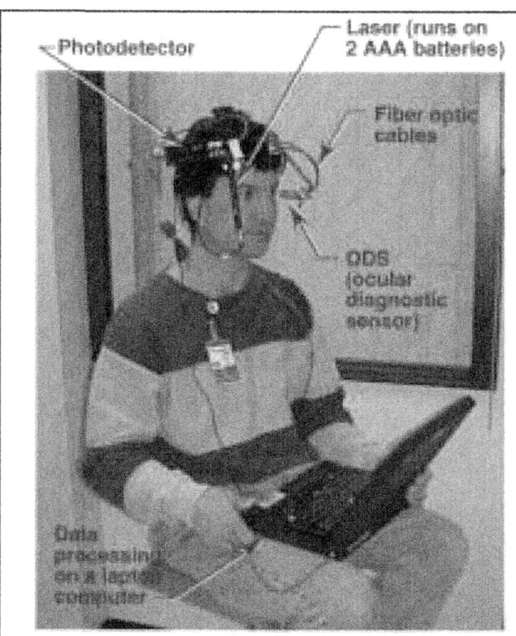

Prototype head mounted eye disease monitoring system.

Glenn Research Center's Dr. Rafat Ansari developed this prototype head-mounted eye disease monitoring system to detect eye cataracts and other eye diseases at the molecular level.

man-made disasters is currently being field tested by the Maryland Emergency Management Agency. Developed by Aeptec Microsystems, Inc., through funding from Goddard's **Small Business Innovation Research** program and the U.S. Federal Emergency Management Agency (FEMA), Earth Alert will make it easier for FEMA, fire departments, and other organizations to track and communicate with emergency vehicles and staff responding to disasters. Earth Alert combines global positioning satellite communications with personal digital assistants and cellular phone technology to effectively integrate, analyze, and disseminate information for emergency management. The technology employs seamless moveable maps and decision-making notification software while providing dispatchers with a real-time location of personnel from the field. It is envisioned that Earth Alert-equipped pagers and fixed receivers located in schools, hospitals, businesses, and other facilities will be able to receive critical notifications of tornadoes, floods, chemical spills, and other disasters. NASA holds the rights to distribute Earth Alerts to first responders, while Aeptec is pursuing commercial applications.

Public Safety

From bombings and other homeland security threats, to child abductions and verifying the "real" Saddam Hussein, a video enhancement system developed at Marshall Space Flight Center is proving to be a valuable law enforcement tool. The technology known as VISAR (short for Video Image Stabilization and Registration) can turn dark, jittery images captured by home video, security systems, and video cameras mounted in police cars into clearer, stable images. NASA scientists Dr. David Hathaway and Paul Meyer, who study violent explosions on the Sun and examine hazardous weather conditions on Earth, created VISAR to aid in their space program research. VISAR has been licensed commercially by Intergraph Corp., of Huntsville, Alabama, and incorporated into Video Analyst,™ a workstation that can stabilize and enhance video, brighten dark pictures and enlarge small sections of pictures to reveal clues about crimes. In a recent application, ABC News asked Intergraph to analyze video clips that aired on Iraqi television on March 20, 2003, apparently showing Saddam Hussein. Officials wanted to verify if Hussein survived a U.S. air strike the previous day, or whether the video was that of a body double. Using Video Analyst with VISAR, it took about 90 minutes to compare the ABC footage to prior Iraqi television images of Hussein and determine—with 99 percent certainty—it was Hussein, according to Intergraph officials. Columbia accident investigators also relied on VISAR to enhance video images of the Space Shuttle's external tank as it shed insulation during liftoff.

Health

In the medical field, technology originally developed to study the behavior of fluids in microgravity is now being used to detect various eye problems earlier and more accurately. Dr. Rafat Ansari, biofluid sensor systems scientist at NASA's Glenn Research Center, utilized the "built-for-space" fiber-optic probe, based on a technique called dynamic light scattering, to detect cataracts and other eye diseases at the molecular level. The probe's value in early cataract detection has already been demonstrated in clinical trials at the National Eye Institute of the National Institutes of Health. Detecting cataracts at an early stage can help doctors find non-surgical cures for the disease. With the help of renowned eye researchers around the world, Ansari is testing the probe as a noninvasive diagnostic measurement device for other eye diseases such as glaucoma and macular degeneration. Ansari also uses the probe in tests to monitor diabetes and Alzheimer's disease.

From public safety and health to environmental conservation, NASA outreach efforts are keeping Americans in step with a world constantly affected by change, while helping to unravel the mysteries of the universe and worlds beyond. The intertwining of Earth and space sciences and the continuance of successful partnership relationships with industry and academia will yield infinite benefits for many years to come.

NASA
Partnering for Success
NASA and You
Investing

Technology Transfer Network and Affiliations

NASA's Technology Transfer Network strives to ensure that the Agency's research and development activities reach the widest possible audience with the broadest impact. The network serves as a resource of scientific and technical information with real-world applications for U.S. businesses interested in accessing, utilizing, and commercializing NASA technology.

FY 2003 Technology Transfer Network and Affiliations

The NASA Technology Transfer Partnership program sponsors a number of organizations around the country that are designed to assist U.S. businesses in accessing, utilizing, and commercializing NASA-funded research and technology. These organizations work closely with the Technology Transfer Offices, located at each of the 10 NASA field centers, providing a full range of technology transfer and commercialization services and assistance.

Technology Transfer Network

The National Technology Transfer Center **<http://www.nttc.edu>**, located on the campus of Wheeling Jesuit University in Wheeling, West Virginia, was established by Congress in 1989 to strengthen American industry by providing access to more than $70 billion worth of federally funded research. By helping American companies use Federal technologies, the NTTC helps them manufacture products, create jobs, and foster partnerships between Federal laboratories and the private sector, universities, innovators, and economic development organizations. From that mission, the NTTC has grown into a full-service technology commercialization center. In addition to providing access to Federal technology information, the NTTC provides technology commercialization training; technology assessment services that help guide industries in making key decisions regarding intellectual property and licensing; and assistance in finding strategic business partners and electronic business development services.

The NTTC developed a leads management system for NASA that is the formal reporting and tracking system for partnerships being developed between NASA and U.S. industry. The leads system allows all members of the NASA Technology Commercialization Team to have an easy-to-use and effective tool to create and track leads in order to bring them to partnerships. The NTTC also utilizes the expertise of nationally recognized technology management experts to create and offer technology commercialization training. Course topics range from the basics of technology transfer to hands-on valuation, negotiation, and licensing. Courses are developed at the NTTC and around the country. In addition, online courses, supporting publications, comprehensive software applications, and videotapes are also available.

NASA TechTracS **<http://technology.nasa.gov>** provides access to NASA's technology inventory and numerous examples of the successful transfer of NASA-sponsored technology for commercialization. TechFinder, the main feature of the Internet site, allows users to search technologies and success stories, as well as submit requests for additional information. All NASA field centers submit information to the TechTracS database as a means of tracking technologies that have potential for commercial development.

Since their inception in January 1992, the six NASA-sponsored Regional Technology Transfer Centers (RTTCs) have helped U.S. businesses investigate and utilize NASA and other federally funded technologies for companies seeking new products, improvements to existing products, or solutions to technical problems. The RTTCs provide technical and business assistance to several thousand customers every year.

The network of RTTCs is divided as follows: Far West (AK, AZ, CA, HI, ID, NV, OR, WA): **The Far West Regional Technology Transfer Center (FWRTTC)** **<http://www.usc.edu/dept/engineering/TTC/NASA>** is an engineering research center within the School of Engineering at the University of Southern California in Los Angeles. Using the Remote Information Service to generate information from hundreds of Federal databases, FWRTTC staff work closely with businesses and entrepreneurs to identify opportunities, expertise, and other necessary resources. The FWRTTC enhances the relationships between NASA and the private sector by offering many unique services, such as the NASA Online Resource Workshop, NASA Tech Opps, and links to funding and conference updates.

Mid-Atlantic (DC, DE, MD, PA, VA, WV): The **Technology Commercialization Center (TeCC)** **<http://www.teccenter.org>**, located in Newport News, Virginia, coordinates and assists in the transfer of marketable technologies, primarily from Langley Research Center, to private industry interested in developing and commercializing new products.

Mid-Continent (AR, CO, IA, KS, MO, MT, ND, NE, NM, OK, SD, TX, UT, WY): **The Mid-Continent Technology Transfer Center (MCTTC)** **<http://www.mcttc.com/>**, under the direction of the Technology and Economic Development Division of the Texas Engineering Service, is located in College Station, Texas. The MCTTC, which provides a link between private companies and Federal laboratories, reports directly to the Johnson Space Center. The assistance focuses on high-tech and manufacturing companies that need to acquire and commercialize new technology.

Mid-West (IL, IN, MI, MN, OH, WI): **The Great Lakes Industrial Technology Center (GLITeC)** **<http://www.glitec.org>**, managed by Battelle Memorial Institute, is located in Cleveland, Ohio. GLITeC works with industries primarily within its six-state region to acquire and use NASA technology and expertise, especially at the Glenn Research Center. Each year, over 500 companies work with GLITeC and its affiliates to identify new market and product opportunities. Technology-based problem solving, product planning and development, and technology commercialization assistance are among the services provided.

Northeast (CT, MA, ME, NH, NJ, NY, RI, VT): The **Center for Technology Commercialization (CTC)** <**http://www.ctc.org**> is a nonprofit organization, based in Westborough, Massachusetts. Covering New England, New York, and New Jersey, the CTC currently has eight satellite offices that form strong relationships with Northeast industry. Operated by the CTC, the NASA Business Outreach Office stimulates business among regional contractors, NASA field centers, and NASA prime contractors.

Southeast (AL, FL, GA, KY, LA, MS, NC, SC, TN): The **Southeast Regional Technology Transfer Center (SERTTC)** <**http://www.edi.gatech.edu/nasa**> at the Georgia Institute of Technology facilitates and coordinates private industry interests in the transfer and commercialization of technologies resulting from NASA's space and Earth science research. Assistance is also provided in Small Business Innovation Research and Small Business Technology Transfer applications, as well as the establishment of connections to specialized research needs within NASA research and development centers nationwide.

NASA Incubator Programs

Ten NASA incubators are included within this network of programs. They are designed to nurture new and emerging businesses with the potential to incorporate technology developed by NASA. They offer a wide variety of business and technical support services to increase the success of participating companies.

Ames Technology Commercialization Center (ATCC) <**http://technology.arc.nasa.gov/smallbusiness.html**>, located in San Jose, California, provides opportunities for start-up companies to utilize NASA technologies. The center uses a laboratory-to-market approach that takes the technological output of Ames' laboratories and pairs that technology with appropriate markets to create and foster new industry and jobs. The incubator helps businesses and entrepreneurs find NASA technology with commercial potential, then provides access to a network of business experts in marketing, sales, high-tech management and operations, financing, and patent and corporate law. The ATCC also offers low-cost office space and other start-up services.

BizTech <**http://www.biztech.org**>, of Huntsville, Alabama, is a small business incubator, offering participating companies access to services at Marshall Space Flight Center laboratories for feasibility testing, prototype fabrication, and advice on technology usage and transfer. BizTech is sponsored by the Huntsville-Madison County Chamber of Commerce.

The Emerging Technology Centers (ETC) <**http://www.etcbaltimore.com**>, located in Baltimore, Maryland, is one of the newest NASA-affiliated incubators. Partnering institutions include the Goddard Space Flight Center and area universities and colleges.

The Florida/NASA Business Incubation Center (FNBIC) <**http://www.trda.org/fnbic/**> is a joint partnership of NASA's Kennedy Space Center, Brevard Community College, and the Technological Research and Development Authority. The mission of the FNBIC is to increase the number of successful technology-based small businesses originating in, developing in, or relocating to Brevard County. The FNBIC offers support facilities and programs to train and nurture new entrepreneurs in the establishment and operation of developing ventures based on NASA technology.

The Hampton Roads Technology Incubator (HRTI) <**http://www.hr-incubator.org**> identifies and licenses NASA Langley Research Center technologies for commercial use. The HRTI's mission is to increase the number of successful technology-based companies originating in, developing in, or relocating to the Hampton Roads area.

The Lewis Incubator for Technology (LIFT) <**http://www.liftinc.org**>, managed by Enterprise Development, Inc., provides outstanding resources for technology and support to businesses in the Ohio region. Its primary objectives are to create businesses and jobs in Ohio and to increase the commercial value of NASA knowledge, technology, and expertise. LIFT offers a wide range of services and facilities to the entrepreneur to increase the probability of business success.

The Mississippi Enterprise for Technology <**http://www.mset.org**> is sponsored by NASA and the Mississippi University Consortium and Department of Economic and Community Development, as well as the private sector. The mission of the enterprise is to help small businesses utilize the scientific knowledge and technical expertise at the Stennis Space Center. A significant part of this effort is Stennis' Commercial Remote Sensing Program, which was formed to commercialize remote sensing, geographic information systems, and related imaging technologies.

The NASA Commercialization Center (NCC) <**http://www.nasaincubator.csupomona.edu**>, run by California State Polytechnic University, Pomona, is a business incubator dedicated to helping small businesses access and commercialize Jet Propulsion Laboratory and Dryden Flight Research Center technologies.

The **UH-NASA Technology Commercialization Incubator** <**http://www.research.uh.edu**> is a partnership between NASA's Johnson Space Center and the University of Houston. The incubator is designed to help local small and mid-sized Texas businesses commercialize space technology. The University of Houston houses the program and provides the commercialization and research expertise of its business and engineering faculties.

Other organizations devoted to the transfer of NASA technology are the **Research Triangle Institute (RTI)** <**http://www.rti.org**>, and the **MSU TechLink Center** <**http://techlink.msu.montana.edu**>.

RTI, located in Research Triangle Park, North Carolina, provides a range of technology management services to NASA. RTI performs technology assessments to determine applications and commercial potential of NASA technology, as well as market analysis, and commercialization and partnership development. RTI works closely with all of NASA's Technology Transfer Offices.

The **MSU TechLink Center**, located at Montana State University-Bozeman, was established in 1997 to match the technology needs of client companies with resources throughout NASA and the Federal laboratory system. TechLink focuses on a five-state region that includes Idaho, Montana, North Dakota, South Dakota, and Wyoming. Working closely with public, private, and university programs, TechLink provides ongoing support in the process of adapting, integrating, and commercializing NASA technology.

Affiliated Organizations, Services, and Products

To complement the specialized centers and programs sponsored by the NASA Technology Transfer Partnership program, affiliated organizations and services have been formed to strengthen NASA's commitment to U.S. businesses. Private and public sector enterprises build upon NASA's experience in technology transfer in order to help with the channeling of NASA technology into the commercial marketplace.

The NASA **Small Business Innovation Research (SBIR)** program <**http://www.sbir.nasa.gov**> provides seed money to U.S. small businesses for developing innovative concepts that meet NASA mission requirements. Each year, NASA invites small businesses to offer proposals in response to technical topics listed in the annual SBIR program solicitation. The NASA field centers negotiate and award the contracts, as well as monitor the work.

NASA's SBIR program is implemented in three phases:

- **Phase I** is the opportunity to establish the feasibility and technical merit of a proposed innovation. Selected competitively, NASA Phase I contracts last 6 months and must remain under specific monetary limits.
- **Phase II** is the major research and development effort, which continues the most promising of the Phase I projects based on scientific and technical merit, results of Phase I, expected value to NASA, company capability, and commercial potential. Phase II places greater emphasis on the commercial value of the innovation. The contracts are usually in effect for a period of 24 months and again must not exceed specified monetary limits.
- **Phase III** is the process of completing the development of a product to make it commercially available. While the financial resources needed must be obtained from sources other than the funding set aside for the SBIR, NASA may fund Phase III activities for follow-on development or for production of an innovation for its own use.

The SBIR Management Office, located at the Goddard Space Flight Center, provides overall management and direction of the SBIR program.

The NASA **Small Business Technology Transfer (STTR)** program <**http://www.sbir.nasa.gov**> awards contracts to small businesses for cooperative research and development with a research institution through a uniform, three-phase process. The goal of Congress in establishing the STTR program was to transfer technology developed by universities and Federal laboratories to the marketplace through the entrepreneurship of a small business.

Although modeled after the SBIR program, STTR is a separate activity and is separately funded. The STTR program differs from the SBIR program in that the funding and technical scope is limited and participants must be teams of small businesses and research institutions that will conduct joint research.

The **Federal Laboratory Consortium (FLC) for Technology Transfer** <**http://www.federallabs.org**> was organized in 1974 to promote and strengthen technology transfer nationwide. More than 600 major Federal laboratories and centers, including NASA, are currently members. The mission of the FLC is twofold:

- To promote and facilitate the rapid movement of Federal laboratory research results and technologies into the mainstream of the U.S. economy.
- To use a coordinated program that meets the technology transfer support needs of FLC member laboratories, agencies, and their potential partners in the transfer process.

The **National Robotics Engineering Consortium (NREC)** <**http://www.rec.ri.cmu.edu**> is a cooperative venture among NASA, the city of Pittsburgh, the State of Pennsylvania, and Carnegie Mellon's Robotics Institute.

Its mission is to move NASA-funded robotics technology to industry. Industrial partners join the NREC with the goal of using technology to gain a greater market share, develop new niche markets, or create entirely new markets within their area of expertise.

The road to technology commercialization begins with the basic and applied research results from the work of scientists, engineers, and other technical and management personnel. The NASA **Scientific and Technical Information (STI) Program** <**http://www.sti.nasa.gov**> provides the widest appropriate dissemination of NASA's research results. The STI Program acquires, processes, archives, announces, and disseminates NASA's internal—as well as worldwide—STI.

The NASA STI Program offers users such things as Internet access to its database of over three million abstracts, online ordering of documents, and the NASA STI Help Desk for assistance in accessing STI resources and information. Free registration with the program is available through the NASA Center for AeroSpace Information.

For more than 3 decades, reporting to industry on any new, commercially significant technologies developed in the course of NASA research and development efforts has been accomplished through the publication of ***NASA Tech Briefs*** <**http://www.nasatech.com**>.

The monthly magazine features innovations from NASA, industry partners, and contractors that can be applied to develop new or improved products and solve engineering or manufacturing problems. Authored by the engineers or scientists who performed the original work, the briefs cover a variety of disciplines, including computer software, mechanics, and life sciences. Most briefs offer a free supplemental technical support package, which explains the technology in greater detail and provides contact points for questions or licensing discussions.

Aerospace Technology Innovation <**http://nctn.hq.nasa.gov/innovation/index.html**> is published bimonthly by the NASA Office of Aerospace Technology. Regular features include current news and opportunities in technology transfer and commercialization, aerospace technology and development, and innovative research.

NASA Spinoff <**http://www.sti.nasa.gov/tto/spinoff.html**> is an annual print and online publication featuring current research and development efforts, the NASA Technology Transfer Partnership Program, and successful commercial and industrial applications of NASA technology.

FY 2003 Technology Transfer Network

The FY 2003 NASA Technology Transfer Network (NTTN) extends from coast to coast. For specific information concerning commercial technology activities described below, contact the appropriate personnel at the facilities listed or go to the Internet at: <**http://nctn.hq.nasa.gov**>. General inquiries may be forwarded to the National Technology Transfer Center.

To publish your success about a product or service you may have commercialized using NASA technology, assistance, or know-how, contact the NASA Center for AeroSpace Information or go to the Internet at: <**http://www.sti.nasa.gov/tto/contributor.html**>.

★ **NASA Headquarters** manages the Spinoff Program.

▲ **Field Center Technology Transfer Offices** represent NASA's technology sources and manage center participation in technology transfer activities.

✕ **National Technology Transfer Center (NTTC)** provides national information, referral, and commercialization services for NASA and other government laboratories.

● **Regional Technology Transfer Centers (RTTC)** provide rapid access to information, as well as technical and commercialization services.

■ **Research Triangle Institute (RTI)** provides a range of technology management services including technology assessment, valuation and marketing, market analysis, intellectual property audits, commercialization planning, and the development of partnerships.

FY 2003 Technology Transfer Network

★ **NASA Headquarters**
National Aeronautics and Space Administration
300 E Street, SW
Washington, DC 20546
NASA Spinoff Program Manager
Janelle Turner
Phone: (202) 358-0704
Email: Janelle.B.Turner@nasa.gov

▲ **FIELD CENTERS**

Ames Research Center
National Aeronautics and Space Administration
Moffett Field, California 94035
Chief, Technology Transfer Office:
Carolina Blake
Phone: (650) 604-1754
Email: Carolina.M.Blake@nasa.gov

Dryden Flight Research Center
National Aeronautics and Space Administration
4800 Lilly Drive, Building 4839
Edwards, California 93523-0273
Chief, Public Affairs and Technology Transfer Office:
Jennifer Baer-Riedhart
Phone: (661) 276-3689
Email: Jenny.L.Baer-Riedhart@nasa.gov

John H. Glenn Research Center at Lewis Field
National Aeronautics and Space Administration
21000 Brookpark Road
Cleveland, Ohio 44135
Director, Technology Transfer Office:
Larry Viterna
Phone: (216) 433-3484
Email: Larry.A.Viterna@nasa.gov

Goddard Space Flight Center
National Aeronautics and Space Administration
Greenbelt, Maryland 20771
Chief, Technology Transfer Office:
Nona K. Cheeks
Phone: (301) 286-5810
Email: Nona.K.Cheeks@nasa.gov

Jet Propulsion Laboratory
4800 Oak Grove Drive
Pasadena, California 91109
Manager, Technology Transfer Office:
James K. Wolfenbarger
Phone: (818) 354-2577
Email: James.K.Wolfenbarger@nasa.gov

Lyndon B. Johnson Space Center
National Aeronautics and Space Administration
Houston, Texas 77058
Director, Technology Transfer Office:
Charlene E. Gilbert
Phone: (281) 483-0474
Email: charlene.e.gilbert@nasa.gov

John F. Kennedy Space Center
National Aeronautics and Space Administration
Kennedy Space Center, Florida 32899
Chief, Technology Programs and Commercialization Office:
James A. Aliberti
Phone: (321) 867-6224
Email: Jim.Aliberti@nasa.gov

Langley Research Center
National Aeronautics and Space Administration
Hampton, Virginia 23681-2199
Director, Program Development and Management Office:
Richard T. Buonfigli
Phone: (757) 864-2915
Email: Richard.T.Buonfigli@nasa.gov

George C. Marshall Space Flight Center
National Aeronautics and Space Administration
Marshall Space Flight Center, Alabama 35812
Manager, Technology Transfer Office:
Vernotto C. McMillan
Phone: (256) 544-2615
Email: Vernotto.C.McMillan@nasa.gov

John C. Stennis Space Center
National Aeronautics and Space Administration
Stennis Space Center, Mississippi 39529
Manager, Technology Transfer Office:
Robert C. Bruce
Phone: (228) 688-1646
Email: Robert.C.Bruce@nasa.gov

✕ **NATIONAL TECHNOLOGY TRANSFER CENTER (NTTC)**
Wheeling Jesuit University
Wheeling, West Virginia 26003
Joseph Allen, President
Phone: (304) 243-2455
Email: jallen@nttc.edu

● REGIONAL TECHNOLOGY TRANSFER CENTERS (RTTCs)

Far West
Technology Transfer Center
3716 South Hope Street, Suite 200
Los Angeles, California 90007-4344
Kenneth Dozier, Director
Phone: (800) 642-2872
Email: kdozier@mizar.usc.edu

Mid-Atlantic
Technology Commercialization Center
12050 Jefferson Ave., Suite 340
Newport News, Virginia 23606
Duncan McIver, Director
Phone: (757) 766-9200
Email: dmciver@teccenter.org

Mid-Continent
Technology Transfer Center
Texas Engineering Extension Service
Technology & Economic Development Division
College Station, Texas 77840-7896
Gary Sera, Director
Phone: (979) 845-2907
Email: gary.sera@teexmail.tamu.edu

Mid-West
Great Lakes Industrial Technology Center (GLITeC)
20445 Emerald Parkway Drive, SW, Suite 200
Cleveland, Ohio 44135
Marty Kress, President
Phone: (216) 898-6400
Email: kressm@batelle.org

Northeast
Center for Technology Commercialization (CTC)
1400 Computer Drive
Westboro, Massachusetts 01581
James P. Dunn, Director
Phone: (508) 870-0042
Email: jdunn@ctc.org

Southeast
Technology Transfer Center (SERTTC)
151 6th Street, 216 O'Keefe Building
Atlanta, Georgia 30332
David Bridges, Director
Phone: (404) 894-6786
Email: david.bridges@edi.gatech.edu

■ RESEARCH TRIANGLE INSTITUTE (RTI)
Technology Application Team
3040 Cornwallis, P.O. Box 12194
Research Triangle Park, North Carolina 27709-2194
Dan Winfield, Executive Director
Phone: (919) 541-6431
Email: winfield@rti.org

NASA CENTER FOR AEROSPACE INFORMATION
Spinoff Project Office
NASA Center for AeroSpace Information
7121 Standard Drive
Hanover, Maryland 21076-1320
Jutta Schmidt, Project Manager
Phone: (301) 621-0182
Email: jschmidt@sti.nasa.gov

Michelle Birdsall, Writer/Editor
Jamie Janvier, Writer/Editor
John Jones, Graphic Designer
Deborah Drumheller, Publications Specialist

Printed in Great Britain
by Amazon